Engineering Dielectric Liquid Applications

Engineering Dielectric Liquid Applications

Special Issue Editor

Issouf Fofana

MDPI • Basel • Beijing • Wuhan • Barcelona • Belgrade

MDPI

Special Issue Editor
Issouf Fofana
Université du Québec à Chicoutimi (UQAC)
Canada

Editorial Office
MDPI
St. Alban-Anlage 66
4052 Basel, Switzerland

This is a reprint of articles from the Special Issue published online in the open access journal *Energies* (ISSN 1996-1073) in 2018 (available at: https://www.mdpi.com/journal/energies/special_issues/ eng_dielectr_liq)

For citation purposes, cite each article independently as indicated on the article page online and as indicated below:

LastName, A.A.; LastName, B.B.; LastName, C.C. Article Title. *Journal Name* **Year**, *Article Number*, Page Range.

ISBN 978-3-03897-402-4 (Pbk)
ISBN 978-3-03897-403-1 (PDF)

Contents

About the Special Issue Editor

Issouf Fofana, Professor and IET Fellow, held the Canada Research Chair on insulating liquids and mixed dielectrics for electrotechnology (ISOLIME) from 2005 to 2015. At his university, he serves as Director of the International Research Centre on Atmospheric Icing and Power Network Engineering (CENGIVRE) and is the research chair on the Ageing of Power Network Infrastructure (ViAHT). Dr Fofana is/was a member of a number of technical/scientific committees of international conferences (including IEEE ICDL, IEEE CEIDP, IEEE ICHVE, CATCON, and ISH, etc.), as well as a member of IEEE DEIS and CEIDP AdComs. He is also serving the scientific community as an Associate Editor for IET GTD, IEEE TDEI, Guest Editor of *Energies*, and chair of the IEEE DEIS Technical Committee on Dielectric Liquids. He is also member of a few working groups (CIGRE and ASTM). Dr Fofana's research in the area of HV engineering has focused on insulation systems relevant to power equipment. His lifetime publication record includes more than 270 scientific publications and three patents.

energies

Editorial

Engineering Dielectric Liquid Applications

Issouf Fofana * and U. Mohan Rao

Research Chair on the Aging of Power Network Infrastructure (ViAHT), Université du Québec à Chicoutimi, Chicoutimi, QC G7H 2B1, Canada; mohan13.nith@gmail.com
* Correspondence: ifofana@uqac.ca; Tel.: +1-418-545-5011

Received: 5 September 2018; Accepted: 10 October 2018; Published: 15 October 2018

1. Introduction

Engineering liquid dielectric applications in electrical apparatus is a field of study that mostly revolves around condition monitoring, performance analysis, aging, diagnostic, and prognostic studies. Liquid dielectrics in the power engineering domain are used for filling and impregnation purposes in various important electrical apparatus. The potential functionalities of these liquids include: Providing electrical insulation, acting as a coolant, and serving as a diagnostic medium. Liquid dielectrics in electrical apparatus have a significant history and have made a remarkable journey in serving the electrical power industry. Their effective functioning over the years has ensured reliability and efficiency in the smooth operation of electrical grids and in ensuring safety for the equipment and personnel in operation.

Since the inception of liquid dielectrics, they have been subjected to rigorous research for ensuring the effectiveness in serving their functions to the industry. Over the last few decades, numerous liquids have been used in research, which only a few have survived in terms of serving the intended purpose successfully and consistently. Mineral insulating oils extracted from crude petroleum stock have been put in service for a long while. However, in recent years, mineral oils have been facing industrial and environmental critiques owing to the developments in the high voltage sector and environmental regulations.

Researchers are making extreme efforts in introducing an appropriate replicate for the existing oils, while several researchers are placing emphasis on modifying the existing ones while including nanotechnology and materials research. However, the contemporary scenario pertaining to the field of liquid dielectrics is in a transit phase in shifting the technology to meaningful and better directions, to meet industry's needs. Numerous research records are positive towards the usage of ester-based biodegradable fluids. Pertaining to all these issues, this Special Issue has been organized with the title "*Engineering Dielectric Liquid Applications*", while emphasizing on electrical apparatus.

This Special Issue was focused on theoretical and practical developments, with special emphasis on engineering problems, in using dielectric liquids in electrical equipment. The applications of new fluids and perspectives were also of particular interest. Listed below, among others, are some the topics of interests considered:

- The application and performance of dielectric liquids;
- Electro–hydrodynamic phenomena and related applications;
- Breakdown and pre-breakdown phenomena;
- Electrostatic charging tendency;
- Insulating liquids efficiency improvement by chemical admixtures;
- Nanofluids and synthetic/vegetable dielectric liquids;
- Measurement, monitoring and diagnostic techniques;
- Fundamental investigations and basic properties.

The topics of the Special Issue aimed at improving the knowledge on existing and new insulating oils in satisfying the requirements of the industry while meeting health and safety regulations. This Special Issue also aims to contribute towards gearing up the research in the application of ester dielectric fluids in oil-filled electrical apparatus.

2. An Outlook of the Special Issue

Engineering Liquid Dielectric Applications, a Special Issue from *Energies*, has been successfully organized with the support extended from the Editorial team of the Journal and the MDPI publishing team. The average processing time of the articles was noted to be 32.16 days; this is only possible because of the time allocated by the reviewers in reviewing the articles. We would like to thank all reviewers for their prompt response during reviewing and revising the manuscripts. This Special Issue has received a good response from researchers, with a remarkable geographical distribution of papers. From the 15 submissions, 10 were accepted. The accepted papers include research from different countries, including Canada (2), China (4), the Czech Republic (1), Germany (1), Poland (1), and Spain (1). Working with insulating oils in the laboratory requires a lot of patience and is exhaustive, particularly for aging studies. We would also like to thank all the researchers who made their contribution to this Special Issue.

At the outset, acceptance and rejection of manuscripts is attributable to several criteria, with major factors including the novelty of the work, state of the art of study, and impact of the article on the existing literature. Importantly, the level of innovation, presentation of the work in the manuscript, hypothesis, and interpretations of results are other factors which make an impression on the reviewers and which may determine the inclusion of an article. Manuscripts possessing the aforementioned features will always be treated as successful and accepted for publication, to be shared with the scientific community. The successful and accepted papers in this Special Issue include 9 research articles and 1 review article in the allied areas of engineering liquid dielectrics. The summary of the articles published in this issue is discussed in the subsequent sections of this editorial.

3. A Review of the Special Issue

Qin et al. [1] experimentally investigated the influence of bubbles on the breakdown voltages of transformer oil and oil-impregnated pressboard under 50 Hz alternating current and direct current (DC) voltages while considering cylinder-plan, sphere-plan electrodes, and cone-plan electrodes. It is reported that, under the influence of bubble, the breakdown voltage of the cylinder-plan electrode dropped the most and the breakdown voltage of the cone-plan electrode dropped the least. In DC, the decrease of the breakdown voltages of the cone-plan and sphere-plan electrodes influenced by the bubbles were much less than that in alternative current (AC). Additionally, for the oil-impregnated pressboard, the decrease of the breakdown voltage under DC voltage was more than that under AC voltage.

Articles [2,3] are in the lines of modifying the properties of insulating fluids using suitable nanoparticles. In [2], an attempt has been made by Mentlik et al. to improve the properties of the natural ester available in the Central European region that is a rapeseed (*Brassica napus seed*) oil. Authors tried to modify the properties of natural esters through percolation treatment and oxidation inhibition by a phenolic-type inhibitor and further using titanium dioxide (TiO_2) nanoparticles with a silica surface treatment. Authors reported and enumerated the dielectric properties, including breakdown voltage, the dissipation factor, and resistivity of natural ester by varying the concentration of the nanoparticles. Velasco et al. [3] focused on the comparison of positive streamers in three different systems: Mineral oil, nanofluids, and immersed dielectric solids in mineral oil. The challenges for the simulation of the streamer in liquid dielectrics with finite-element software have been also discussed. It is observed that the dielectric solid blocks the propagation of the streamer when it is submerged with a horizontal orientation, thus perpendicular to the applied electric field.

A mathematical model describing the kinetics of drying according to temperature, initial moisture, paper weight, final moisture, and extraction rate is proposed by Betie et al. [4], based on thermogravity analyses. The impact of moisture, weight, and temperature on the drying process has been investigated. Authors reported a mathematical model to describe the kinetics of drying according to temperature, initial moisture, paper weight, final moisture, and extraction rate. The amount of moisture removed at the end of the drying process has been also demonstrated by the authors, using the proposed model. In article [5], the effect of three different carboxylic acids on the aging of oil/paper insulation used in power transformers has been investigated experimentally. The observations have been correlated to the degree of polymerization. Authors concluded that present diagnostic techniques consisting of monitoring insulation oil conditions based on the total acid number (which is used as reclamation criterion) does not provide a true picture of the transformer condition, since this procedure cannot distinguish between different acid types and their influences.

Experimental investigations dedicated towards the performance of natural esters have been presented in [6,7]. Haegele et al. [6] reported the degree of inhomogeneity differences in breakdown voltage between natural ester and mineral oils while using lightning impulses at different electrode arrangements representing different fields in homogeneity factors and different gap distances. Authors also envisaged different electrode and conductor arrangements reported in the literature for oil breakdown voltage studies. Wang et al. [7] discussed the mechanisms characterizing fast and slow streamers in vegetable oil and mineral oil, based on their calculations. The electronic properties of typical molecules were calculated using the quantum chemistry method (B3LYP/6-31G*), based on the density function theory. It was reported that the insulation characteristics of triolein and tristearin are more likely to be degraded under an external electric field than those of 1-methylnaphthalene and eicosane.

Rozga et al. [8] presented the results of the studies on negative streamer propagation in a point-to-sphere electrode system with a pressboard barrier placed between them. Experimental investigations have been carried out on two synthetic esters and two natural esters in comparison to mineral oils. Based on the study, it was reported that the intensity of the discharge processes, comparing the same voltage levels, was mostly higher when streamers developed in ester liquids. Zhang et al. [9] studied the performance of new alternative insulating oil/paper systems based on ester/Nomex, respectively, for high temperature applications, based on a typical loading curve on the China Southern Power Grid. Authors evaluated the physiochemical and mechanical properties along with the thermal index of the oil paper insulation system and reported that the considered new oil/paper insulation system is a promising one for high thermal applications compared to mineral oil/kraft paper insulation systems.

As discussed earlier, this Special Issue also reports on a comprehensive survey by Wang et al. [10], considering 132 citations on existing and new insulating oils for transformer insulation technology. In this survey, authors emphasized and highlighted the basis properties, variation in electrical properties with aging, nanomodified insulating oils, and recent studies on the performance of mineral oil (molecular and atomic levels). Finally, future research hotspots and notable research topics are also discussed for the benefit of researchers.

4. Closing Remarks

Insulating materials are still the Achilles' heel of power equipment. When this type of equipment fails, the cause can generally be traced to insulation degradation/aging. A wide variety of types of insulation is available, spanning all three forms of matter (solid, liquid, and gas), with sometimes a single form involved, but often a combination of forms, such as the solid/liquid or solid/gas forms. Liquid dielectrics are used in a wide range of power equipment, including transformers, reactors, capacitors, etc. The contributions in this Special Issue discuss a wide range of liquid dielectric applications relevant to engineering. Even though very interesting results have been reported, there are still many challenges to be solved in order to meet industrial requirements. Fundamental studies are still needed to improve our basic understanding of the mechanisms involved in insulation system degradation and biodegradable fluid applications in various power components connected to power grids.

In the actual grid components, the insulation system was mostly designed to operate at normal power frequency voltages, with occasional lightning and/or switching transients. However, in today's grids, with the large penetration of renewable energy sources/plug-in vehicles, the reliability of insulation systems has to be improved. Nanotechnology is poised to play an important role in the insulating fluid industry by enhancing the physicochemical properties.

Author Contributions: The authors contributed equally to this work.

Funding: This research received no external funding.

Acknowledgments: Issouf Fofana is grateful to the MDPI Publisher for the invitation to act as guest editor of this special issue and is indebted to the editorial staff of *"Energies"* for the kind cooperation, patience and committed engagement. The guest editor would also like to thank the authors for submitting their excellent contributions to this special issue. Thanks are also extended to the reviewers for evaluating the manuscripts and providing helpful suggestions. U. Mohan Rao, is thankful to Issouf Fofana, for providing an opportunity to co-author this editorial.

Conflicts of Interest: The authors declare no conflict of interest.

References

1. Qin, C.; He, Y.; Shi, B.; Zhao, T.; Lv, F.; Cheng, X. Experimental Study on Breakdown Characteristics of Transformer Oil Influenced by Bubbles. *Energies* **2018**, *11*, 634. [CrossRef]
2. Mentlik, V.; Trnka, P.; Hornak, J.; Totzauer, P. Development of a Biodegradable Electro-Insulating Liquid and Its Subsequent Modification by Nanoparticles. *Energies* **2018**, *11*, 508. [CrossRef]
3. Velasco, J.; Frascella, R.; Albarracín, R.; Burgos, J.C.; Dong, M.; Ren, M.; Yang, L. Comparison of Positive Streamers in Liquid Dielectrics with and without Nanoparticles Simulated with Finite-Element Software. *Energies* **2018**, *11*, 361. [CrossRef]
4. Betie, A.; Meghnefi, F.; Fofana, I.; Yeo, Z. Modeling the Insulation Paper Drying Process from Thermogravimetric Analyses. *Energies* **2018**, *11*, 517. [CrossRef]
5. Kouassi, K.D.; Fofana, I.; Cissé, L.; Hadjadj, Y.; Yapi, K.M.L.; Diby, K.A. Impact of Low Molecular Weight Acids on Oil Impregnated Paper Insulation Degradation. *Energies* **2018**, *11*, 1465. [CrossRef]
6. Haegele, S.; Vahidi, F.; Tenbohlen, S.; Rapp, K.J.; Sbravati, A. Lightning Impulse Withstand of Natural Ester Liquid. *Energies* **2018**, *11*, 1964. [CrossRef]
7. Wang, Y.; Wang, F.; Li, J.; Liang, S.; Zhou, J. Electronic Properties of Typical Molecules and the Discharge Mechanism of Vegetable and Mineral Insulating Oils. *Energies* **2018**, *11*, 523. [CrossRef]
8. Rozga, P.; Stanek, M.; Pasternak, B. Characteristics of Negative Streamer Development in Ester Liquids and Mineral Oil in a Point-To-Sphere Electrode System with a Pressboard Barrier. *Energies* **2018**, *11*, 1088. [CrossRef]
9. Zhang, X.; Ren, L.; Yu, H.; Xu, Y.; Lei, Q.; Li, X.; Han, B. Dual-Temperature Evaluation of a High-Temperature Insulation System for Liquid-Immersed Transformer. *Energies* **2018**, *11*, 1957. [CrossRef]
10. Wang, X.; Tang, C.; Huang, B.; Hao, J.; Chen, G. Review of Research Progress on the Electrical Properties and Modification of Mineral Insulating Oils Used in Power Transformers. *Energies* **2018**, *11*, 487. [CrossRef]

Review

Review of Research Progress on the Electrical Properties and Modification of Mineral Insulating Oils Used in Power Transformers

Xiaobo Wang [1], Chao Tang [1,*], Bo Huang [2], Jian Hao [3] and George Chen [2]

[1]   College of Engineering and Technology, Southwest University, Chongqing 400715, China;
      xiaobo@email.swu.edu.cn
[2]   School of Electronics and Computer Science, University of Southampton, Highfield Campus,
      Southampton SO17 1BJ, UK; bh2e13@ecs.soton.ac.uk (B.H.); gc@ecs.soton.ac.uk (G.C.)
[3]   Laboratory of Power Transmission Equipment & System Security and New Technology,
      Chongqing University, Chongqing 400044, China; cquhaojian@126.com
*   Correspondence: tangchao_1981@163.com; Tel./Fax: +86-023-6825-1265

Received: 29 January 2018; Accepted: 19 February 2018; Published: 26 February 2018

Abstract: In November 2017, the first ± 1100 kV high-voltage direct-current power transformer in the world, which was made by Siemens in Nurnberg, passed its type test. Meanwhile, in early 2017, a ± 1000 kV ultra-high voltage (UHV) substation was officially put into operation in Tianjin, China. These examples illustrate that the era of UHV power transmission is coming. With the rapid increase in power transmission voltage, the performance requirements for the insulation of power transformers are getting higher and higher. The traditional mineral oils used inside power transformers as insulating and cooling agents are thus facing a serious challenge to meet these requirements. In this review, the basic properties of traditional mineral insulating oil are first introduced. Then, the variation of electrical properties such as breakdown strength, permittivity, and conductivity during transformer operation and aging is summarized. Next, the modification of mineral insulating oil is investigated with a focus on the influence of nanoparticles on the electrical properties of nano-modified insulating oil. Recent studies on the performance of mineral oil at molecular and atomic levels by molecular dynamics simulations are then described. Finally, future research hotspots and notable research topics are discussed.

Keywords: mineral insulating oil; electrical properties; power transformer; molecular dynamics simulation

1. Introduction

Insulating oil is widely used as an insulating medium in oil-immersed transformers, allowing the safe and reliable transmission of electrical energy. The first insulating oil applied to transformers was mineral oil extracted from petroleum [1]. Since the 1940s, mineral insulating oil has been widely used in power equipment like transformers, capacitors, cables, and bushings [2].

With the development of science and technology, more and more types of transformer oil have been developed, but mineral insulating oil is still used widely because of its low cost, good insulating properties, low condensation point, and low viscosity. At present, the main raw materials of mineral insulating oil are paraffin-based and naphthenic-based crude oils [3], which are composed of hydrocarbons. The hydrocarbon components include alkanes, cycloalkanes (one-, two-, three-, and four-membered rings), and aromatic hydrocarbons. The two main components are alkanes and cycloalkanes [4], and the only difference between crude oils is the proportion of each component. The specific charts showing the classification and an introduction about different types of insulating oils can be found in reference [2].

1.1. Characteristics of Naphthenic-Based Mineral Insulating Oil

The molecular structures of the main components of the naphthenic-based oils are shown in Figure 1. Important features of naphthenic-based mineral insulating oil are summarized as follows.

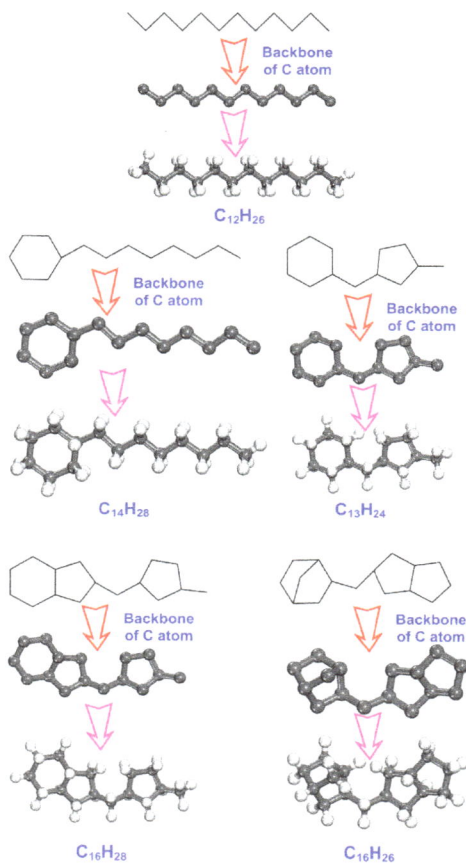

Figure 1. Main components of naphthenic mineral insulating oil.

1.1.1. Suitable Solubility

Naphthenic-based insulating oil has moderate solubility, and can not only dissolve the oil sludge generated from the compound actions of high temperature, electric field, moisture, and metal catalysts, but also prevent the insulating varnish of the transformer from dissolving [5]. Because the naphthenic-base oil can dissolve the oil sludge, it prevents the oil sludge from sticking onto the insulating material and depositing on the circulating oil ducts and cooling fins, helping to avoid the local overheating of the transformer winding and the rise of the transformer operation temperature, prolonging the service life of the transformer.

1.1.2. Good Low-Temperature Properties

The low-temperature environment suitable for the use of transformer oil can be estimated from its pour point. Because paraffinic-based oil contains more paraffin hydrocarbons, which readily crystallize at low temperature, than naphthenic mineral insulating oil, the fluidity of the oil decreases. If the pour point of paraffinic-based oil is lowered through dewaxing, its cost is comparatively high, and the pour point cannot be very low because of the restriction of degree of dewaxing. Compared with that of paraffinic-based oil, the content of paraffin hydrocarbons in naphthenic-based transformer oil is low, and its pour point is low without dewaxing [3]. Therefore, naphthenic-based transformer oil is endowed with good low-temperature properties. In the case of an extreme climate with temperatures as low as −40 °C, naphthenic-based transformer oil still works normally without affecting the insulation properties of the equipment.

1.1.3. Favorable Heat Dissipation

Generally, the high-temperature viscosity of transformer oil needs to be as low as possible to maximize the fluidity and aid heat emission. Some research data show that the kinematic viscosities of naphthenic-based and paraffin-based transformer oils at 40 °C are similar. However, when the temperature reaches 100 °C, the kinematic viscosity of naphthenic-based transformer oil is obviously lower than that of the paraffin-based transformer oil. Therefore, when naphthenic-based oil is used, the heat dissipation and cooling properties of the transformer will be better. QS2598A paraffinic-based oil and V-35 standard naphthenic-based oil have been researched [6]. It was found that the viscosity of naphthenic oil is much lower than that of paraffin-based crude oil when the temperature is −20–50 °C. This shows that normal starting with naphthenic transformer oil is easier than that with paraffinic transformer oil in winter after stopping [3]. Although naphthenic crude oil possesses a variety of good performance parameters, naphthenic-based crude oil is a rare resource and only accounts for 2–3% of the total amount of crude oil [7]. Since the petroleum crisis in the 1970s, the naphthenic-based crude oil resource has gradually decreased [8]. In the 1980s, production of naphthenic-based crude oil rapidly lowered to only 20% of that in 1970s [9].

1.2. Characteristics of Paraffinic-Based Transformer Oil

The molecular structures of the main components of paraffin-based oil are shown in Figure 2. Important features of paraffin-based transformer oils are summarized as follows.

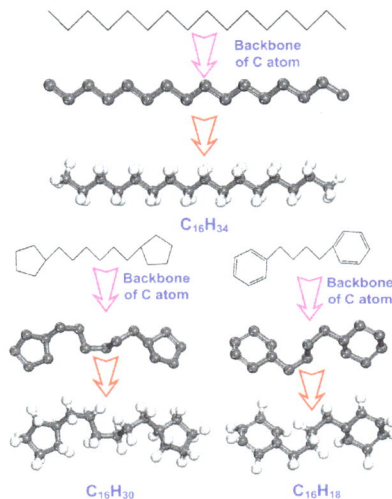

Figure 2. Main components of paraffin-based transformer oil.

1.2.1. Suitable Density

When a transformer operates at extremely low temperature, liquid water will be generated when any ice melts, and the breakdown voltage will be markedly lowered if liquid water flows into the electrode region, so the emergence of floating ice should be prevented as much as possible. Data have shown that the actual density of pure ice changes within the range of 880–920 kg/m^3 at 0 °C and 0.1 MPa [10], therefore, the large difference between densities of the transformer oil and floating ice can readily control the emergence of ice. The density of paraffinic oil is lower than that of naphthenic oil at 0 and 20 °C [3], so paraffinic-based oil is more effective at controlling the emergence of floating ice than naphthenic-based oil.

1.2.2. Favorable Electrical Properties

Electrical performance is an important performance index of transformer oil, and mainly includes the breakdown voltage and dielectric loss. These two parameters are mostly affected by the moisture content of the transformer oil, because even a small amount of moisture strongly influences the breakdown voltage and dielectric loss. The breakdown voltages and dielectric losses of paraffinic-based and naphthenic-based oils are almost identical, and there is no obvious difference between their electrical properties under anhydrous conditions [3,11].

1.2.3. High Antioxidation Stability

Antioxidation stability is an important parameter reflecting the oxidation resistance of transformer oil. The antioxidation stability of paraffinic-based oil is higher than that of naphthenic-based oil [3,12], which means that paraffinic-based oil has a longer service life during long-term operation than naphthenic-based oil.

The density and antioxidation stability of paraffin-based transformer oil are better than those of naphthenic-based transformer oil. However, the solubility, low-temperature performance, and kinematic viscosity of naphthenic-based transformer oil are better than those of paraffin-based transformer oil. Meanwhile, there is little difference between the electrical properties of both oil types. Overall, these two kinds of transformer oil have their own advantages and disadvantages. However, naphthenic-based oil has superior low-temperature properties, a reasonable proportion of alkanes, cycloalkanes, and arenes, low wax content (generally below 3%), and a complex, expensive dewaxing process is not required. Thus, transformer insulating oil [3] has mainly been refined from naphthenic-based crude oil.

In 2014, the State Grid Corporation of China started the ultra-high voltage (UHV) projects with four alternating current (AC) lines and six direct current (DC) lines (ten lines in all) with a total investment of RMB 220 billion Yuan (about 35 billion US dollars) and total length of 16,000 km. This project passes through 16 provinces and creates a new record in the history of global power construction. China, along with the rest of the world, is rapidly entering a UHV era. As the properties of insulating oil and its aging by-product were proved to have close relation to the operation life of power transformers [13–17]. The continuous rise of the voltage level of power grids and continuous increase of loads mean that insulating materials now face unprecedented challenges.

Therefore, for power transformer oil to meet the insulation performance requirements for UHV power transmission, scholars all over the globe are conducting numerous studies on the modification of transformer insulating oil. In this review, studies of naphthenic mineral insulating oil for power transformers conducted in recent decades are analyzed, and some perspectives on the new developments of mineral insulating oil are discussed.

2. Properties of Mineral Insulating Oil

In a mineral oil-immersed transformer, the insulation system of the transformer consists of mineral insulating oil and insulating paper. In normal operation of such a transformer, the oil-paper insulation system can be influenced by factors such as the electric field, thermal field, and force field. The physicochemical properties of the oil-paper insulation system gradually deteriorate over time. Therefore, these parameters can directly reflect the electrical properties of mineral insulating oil, such as the breakdown strength, dielectric constant, and conductivity. The changes of the above parameters of mineral insulating oil during operation of the transformer are analyzed in this section.

2.1. Breakdown Voltage of Mineral Insulating Oil

The electrical properties and thermal stability of insulating materials are closely correlated, and the electrical properties will deteriorate as the thermal stability of the insulating material decreases. Therefore, it is necessary to study the thermal stability of oil-paper insulation when researching its breakdown voltage. Moisture and temperature both strongly influence the aging process of the oil-paper insulation of transformers. Moisture severely affects the electrical properties of oil-paper insulation, accelerating its aging and shortening its service life [18–20].

For years, a large number of scholars have investigated the formation pathway of water in transformers and its influence on transformer performance. There are basically three states of moisture in insulating liquids, which are the dissolved state, emulsified state, and dispersed state. The moisture content in insulating oil can largely influence the electrical properties, that is, increase the electrical conductivity and dissipation factor and lower the electric strength of transformer oil [21].

In 2004, Liu [22] studied the source of moisture in the transformer, evaluated the amount of water generated through aging of the oil-paper insulation system, and derived a mathematical expression for the relationship between the aging life and moisture content of oil-paper insulation. Chen [23] developed an online calculation model to monitor the moisture content of mineral insulating oil based on the neural network for online monitoring. This model can be used to determine not only the change of the moisture content of the transformer, but also can obtain the cause of abnormal changes of moisture content of mineral insulating oil.

In 2005, Yang [24] researched the identification of the aging degree of oil-paper insulating materials by multivariate analysis of the statistics for diagnosis of the aging of oil-paper insulation. The aging characteristics were measured, such as the degree of polymerization of insulation paper, furfural content of oil, CO and CO_2 contents of mineral insulating oil, acid value of mineral insulating oil, and moisture content. The aging properties of different insulation combinations were systematically compared and analyzed, and the correlation between the aging results for different insulating materials were explored. This research showed that the insulation state of new samples can be reasonably judged using the discrimination function.

In 2012, Liao [25] studied the effect of water on the thermal aging rate and characteristic parameters of mineral oil-paper insulation. It was found that the moisture content and its variation affected the furfural and acid contents in the insulation during the aging process. These results indicated that the influence of water should be taken into account in the condition assessment and fault diagnosis of mineral oil-paper insulation according to the traditional aging characteristic parameters of oil, such as furfural and acid content.

Temperature and moisture synergistically act on the mineral oil-paper insulation system of a transformer. Zhou et al. [26] derived an equation for moisture diffusion in mineral oil-paper insulation considering Fick's second law of diffusion, and established relevant models to characterize the relationship between the aging state of mineral oil-paper insulation and the moisture content. They then studied the effect of various moisture contents on mineral oil-paper insulation at different temperatures. Distribution curves of the change of the moisture concentration with the thickness of the insulating paper at different times were measured, and the average moisture concentration in insulation paper was obtained through an integral computation of these distribution curves. The relationship between the breakdown voltage and water content of mineral insulating oil is shown in Figure 3.

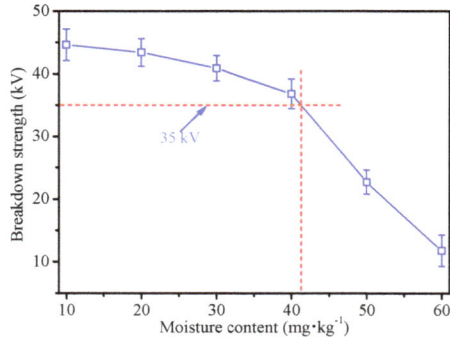

Figure 3. Relationship between the breakdown strength and moisture content of mineral insulating oil.

The thermal aging rate of mineral oil-paper insulation with different temperatures and moisture contents has been studied [25]. A strong positive correlation was found for the fluctuation amplitudes of moisture in oil and moisture in paper, and the relationship between overall fluctuation trends of these two parameters was consistent with the initial moisture content. The moisture content and its changes can affect the variation of the contents of furfural, acid, and other products during aging. When accelerated thermal aging was carried out for oil-paper insulation samples at different temperatures, it was found that the oil type was the main factor affecting the moisture content of the mineral oil.

When the mineral oil-paper insulation system is affected by moisture, the insulating material will degrade over time, the dielectric loss of the oil paper will be increased, the insulation resistance will be lowered, and the operation life of the equipment will be seriously affected. As the moisture content of mineral oil increases, its breakdown voltage rapidly decreases. Over time, the breakdown voltage of the insulating oil will lower, and especially after medium-term aging, the rate of the breakdown voltage decrease accelerates [27–33].

The power frequency breakdown voltage of the Karamay No. 25 transformer oil from the China Petroleum Chemical Co. was tested according to a standard insulation oil breakdown voltage measurement method [34]. This test revealed that the breakdown voltage of No. 25 transformer oil decreased obviously with increasing water content of the transformer oil. When the water content was more than 40 mg/kg, the breakdown voltage of the Karamay No. 25 transformer oil was close to 35 kV, which is the minimum national standard requirement for transformer oil in China.

Besides moisture, the acid generated during aging of insulating paper will enter into the insulating oil, and affect its properties. The acid has a negative influence on the safe operation of the transformer. The acid in the transformer oil may erode the metal parts in the transformer, and then the new material generated can accelerate the oxidation of the insulating oil [35,36].

Research has shown that oil type is the main factor affecting the acid content during oil aging, and the acid content of a mixture of vegetable and mineral oils is greater than that of ordinary transformer oil. When plenty of acid and water have accumulated in an insulation system, a synergistic effect of acid and water accelerates the aging of mineral oil-paper insulation. The accelerated aging effect of the acid on the insulating paper is more obvious for acids with lower molecular weight. Therefore, the acid content of the mineral oil-paper insulation is an important indicator to judge whether the operation status of a transformer is normal or not.

Based on the above analysis, on one hand, the properties of mineral insulating oil are degraded through oxidative degradation over time. On the other hand, moisture, small acid molecules, and gases like CO and CO_2 generated during thermal aging of the insulating paper will enter the mineral insulating oil [37–40]. The structures of water, gas, and acid dissolved in the oil are shown in Figure 4. Moisture and impurities in the oil can form a "small bridge" in a certain direction under the action of the electric field. When the positive and negative electrodes are connected with the small bridge, the leakage current flowing through the small bridge will increase because of the high electrical conductivity of water and impurities, which will lead to local overheating, vaporization of water, and formation of air bubbles. Gases have lower relative dielectric constants and higher withstand voltages than those of liquids, so electrical discharge will occur first in the gas in air bubbles. More gas will be decomposed when the charged particles collide with the oil molecules, the bubble volume will continue to increase, and the bubbles will be arranged to form small air bridges that connect the electrodes under the action of an electric field. Oil breakdown occurs at the moment when the bridge breaks through both electrodes [41–43].

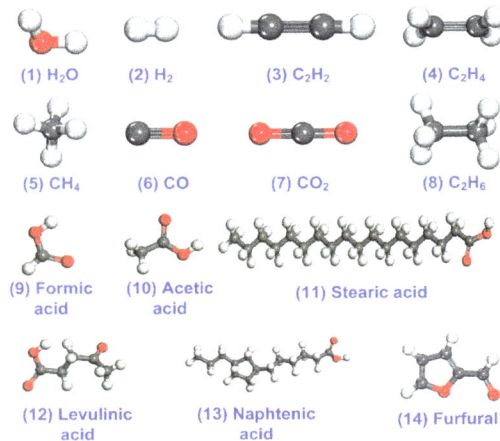

(1) H$_2$O   (2) H$_2$ (3) C$_2$H$_2$ (4) C$_2$H$_4$

(5) CH$_4$ (6) CO   (7) CO$_2$ (8) C$_2$H$_6$

(9) Formic acid   (10) Acetic acid   (11) Stearic acid

(12) Levulinic acid   (13) Naphtenic acid   (14) Furfural

Figure 4. The water, gases, acids, etc. dissolved in mineral insulating oil of transformer.

2.2. Dielectric Constant of Mineral Insulating Oil

During the actual operation of mineral insulating oil, the molecules of the mineral insulating oil undergo physical and chemical reactions because of the influence of light, electricity, and magnetism, to form a series of products, which strongly affect the dielectric properties of the mineral insulating oil. Experiments using the return voltage measurement method [44], depolarizing current method [45,46], and frequency domain dielectric spectroscopy [47–49] have revealed that the free radical chain reaction process of autoxidation will occur in the mineral insulating oil during aging, which increases the acid content of the mineral insulating oil. Resinous substances, oligomers, and viscous substances with high molecular weight will be generated through further condensation reactions between acid and alcohol species. The moisture content will increase with the amount of compounds generated by this process, eventually leading to increased contents of oxides, alcohols, aldehydes, ketones, and acidic compounds along with water. These species intensify intermolecular thermal motion; increase the number of equivalent dipoles; disconnect molecular chains; weaken the cross-linking force; increase the polarization capacity of mineral insulating oil [50]; increase kinematic viscosity, dielectric constant, and the dielectric dissipation factor, and decrease recovery voltage. The change of these parameters affects the dielectric properties of the mineral insulating oil.

Zhou et al. [51] developed a modified Coelho model, which introduced the current density in the external circuit to modify the space charge polarization theory of Coelho. It brings a better understanding of the low-frequency response in mineral insulating oil. It can also be used to explain the dielectric behavior of traditional mineral insulating oil.

2.3. Electrical Conductivity of Mineral Insulating Oil

Zaengl [52] and Saha [53] conducted the first research on the electrical conductivity of mineral insulating oil. Their tests showed that the higher the electrical conductivity of mineral insulating oil, the higher the initial value of polarization current, and the lower the electrical conductivity of mineral insulating oil, the smaller the initial current value, which can considerably prolong the current time decayed by index. In 2010, Ma [54] and Yang [55] measured the DC electrical conductivity of mineral insulating oil at 90 °C according to the standard 'Determination of Volume Resistivity of Power Oils' [56]. They found that the DC electrical conductivity of the mineral insulating oil gradually increased during the aging process. According to their analysis, this was caused by the dissolution of the organic acids and other substances generated through oxidative degradation of the mineral insulating oil during the aging process. By analyzing the change of the acid content of mineral insulating oil, Hao [49] concluded that an increased acid content of mineral insulating oil will raise the electrical conductivity of oil products and decrease the insulation properties of the oil. During aging, if the acid content of mineral insulating oil gradually increases, the electrical conductivity also rises, especially during the later stage of aging. A large number of reports indicate that the conductivity of oil increases during the aging of mineral oil-paper insulation [49,54–57].

Zhou and co-workers suggested to the characterize charge carriers in mineral insulating oil by using the polarization and depolarization currents, and simulate the frequency response by the calculated insulating oil conductivity [58,59].

The increase of the electrical conductivity of mineral insulating oil over time is mainly caused by the increasing content of aging products dissolved in the mineral insulating oil, such as water, acid, and furan compounds, during aging [60]. In addition, small particles, such as electrically insulating particles (silica and paper), semiconducting particles (carbon), and conducting particles (copper), have been found to have a direct relationship with the electrical conductivity of mineral insulating oil [61–63]. For example, semiconducting particles such as carbon led to markedly increased conductivity.

Our team measured the relationship between the acid content of mineral insulating oil and electrical conductivity of the mineral oil-paper insulating system. The experimental samples were composed of No. 25 transformer oil produced in Karamay, Xinjiang, and the insulation winding was provided by Chongqing ABB Transformer Co., Ltd. (Chongqing, China) Both sides of the copper strip of the winding were covered with ten layers of cellulose insulating paper with a thickness of 75 μm, and the length and width of the winding were 12 and 2.8 cm, respectively.

The results of the experiment are presented in Figure 5. During the aging process of the mineral oil-paper insulating system, the oleic acid content and electrical conductivity of the mineral insulating oil increased, especially during the later stage of aging. The increase of the acid content of mineral insulating oil will raise its conductivity and degrade its insulation performance.

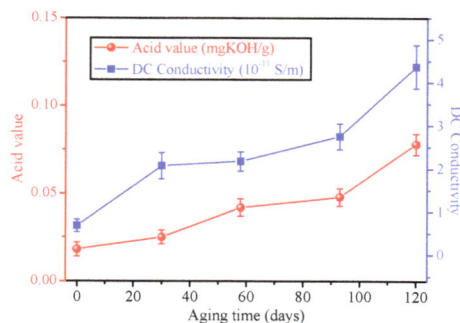

Figure 5. Relationships between the acid value, DC conductivity, and aging time of mineral insulating oil.

3. Modification of Mineral Insulating Oil

The above analysis reveals that temperature, moisture, and small acid molecules are the most important factors affecting the electrical performance of mineral insulating oil. Next, methods to improve the performance of mineral insulating oil based on the main factors leading to the decrease of the properties of mineral insulating oil are introduced.

3.1. Modification with Nanoparticles

In 1994, the concept of a "nano-dielectric" was proposed by Lewis [64], which considers that the properties of nanoscale dielectrics are determined by the interface between the nanoscale dielectric and substrate material. The interface effect is an important characteristic of a nanoscale dielectric that determines its electrical properties. In 1995, the concept of nanofluids was proposed by Choi [65]. Nanoscale additives do not easily settle in a liquid medium, their surface area is large, and their thermal conductivity is high. As a result, these studies opened a new chapter in the modification of mineral insulating oil. Nanoparticles are now widely used to modify mineral insulating oil and represent a new type of material.

3.1.1. Effect of Nanoparticles on the Breakdown Voltage of Mineral Insulating Oil

The decrease of the breakdown voltage is one of the most prominent problems during the aging of mineral insulating oil. Dariusz and Aksamit [66,67] modified new and aged mineral insulating oil with fullerene (C_{60}) (Figure 6), and then measured the breakdown voltage of the modified mineral insulating oil samples. They found that the dielectric loss of the modified oil was lower, and its breakdown voltage remained higher during the aging process.

The water content measurements of the mineral insulating oil samples showed that all the C_{60}-doped samples had lower water contents than those of the samples without added C_{60}. With increasing C_{60} content, the water absorption by the mineral insulating oil samples decreased.

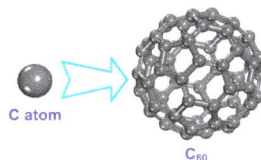

Figure 6. Molecular structure of fullerene (C_{60}).

Prasath et al. [68] modified mineral insulating oil with $CaCu_3Ti_4O_{12}$ (CCTO) nanoparticles, which have a high dielectric constant. Nano-fluids containing mineral oils with of different CCTO contents were prepared using the ultrasonic effect. Important parameters of these nanomaterial-modified mineral insulating oil samples (also called nanofluids) were measured according to the procedures of ASTM (American Society for Testing and Materials) and IEC (International Electrotechnical Commission) standards. The results showed that the AC breakdown voltage increased with the content of CCTO nanoparticles in the mineral insulating oil. The breakdown voltages of mineral insulating oil modified with C_{60} and CCTO nanoparticles are listed in Table 1.

As a semiconductor material, nano-TiO_2 is widely used in the modified mineral insulating oil, thereby increasing the breakdown voltage of the mineral insulating oil, and the molecular model of nano-TiO_2 is shown in Figure 7. When the mineral insulating oil is modified the nano-TiO_2 of the content 1% [69], the value of breakdown voltage of nano-modified mineral insulating oil rises by 1.15 and 1.43 times. For the lightning breakdown voltage, the nano-modified oil was 13.3 kV higher than that of the mineral insulating oil. On this basis, the surfaces of the nano-TiO_2 shall be modified with octadecanoic acid [70], and then added into the mineral insulating oil, which can improve the dialectical property of insulating oil and the breakdown voltage in AC state.

Table 1. Breakdown voltages of mineral insulation oil modified with C_{60} and CCTO.

Nanoparticles	Year, Author	Manufacture Factory, Modified Object	Measurement Standard	Optimal Nanoparticles Concentration	Breakdown Voltage (kV)
C_{60}	2014, Dariusz Z, et al. [66,67]	ORLEN OIL TRAF, fresh uninhibited mineral oil	AC, IEC 60156, spherical spark gap, 2.5 mm gap	250 mg/L 150 mg/L 300 mg/L	About 77 (aged at 110 °C, 0 h) About 67 (aged at 110 °C, 48 h) About 56 (aged at 110 °C, 144 h)
$CaCu_3Ti_4O_{12}$ (CCTO)	2017, Prasath RTAR, et al. [68]	Local manufacturers, mineral oil	AC, IEC 60156, spherical electrode, 2.5 mm gap	0.01% (volume fractions) 0.005% (volume fractions)	46.4 (unaged) 36.7 (ageing based on ASTM D1934)

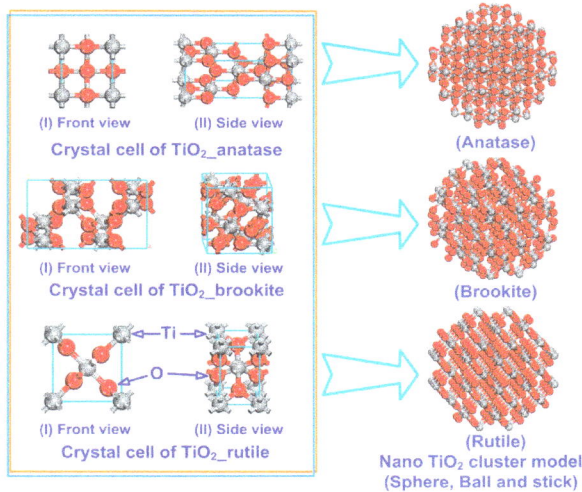

Figure 7. The molecular model of nano-TiO_2.

Wang and Raffia [71] modified mineral insulating oil with 5–40 wt % nano-TiO_2, and measured the impulse breakdown voltage of the oil samples before and after modification according to an IEC standard. Under positive pulse conditions, at a certain concentration, the breakdown strength increased to the maximum value firstly, and then, it began to decrease. Under negative pulse conditions, the breakdown voltage of the samples with different nano-TiO_2 concentrations was lower than that of the unmodified mineral insulating oil. This indicates that when the content of nanoparticles is too high, they will aggregate in the mineral insulating oil, which causes degradation of performance.

The mechanism behind the breakdown voltage increase of mineral insulating oil after nano-TiO_2 modification is that the polarization of nano-TiO_2 acts as an electronic trap in the electron transfer process of electrically stressed nanofluids. In addition, the high specific surface area of the nanoparticles have the effects of increasing the possibility of electron scattering, lowering the electronic impact energy, and preventing oil ionization. The oil diffusion behavior and trap network are changed by the addition of nano-TiO_2, effectively lowering the mobility of carriers.

Du et al. [72] also studied the effect of nano-TiO_2 on the aging performance of the mineral insulating oil. They carried out accelerated aging testing of unmodified and nano-TiO_2-modified mineral insulating oil samples for 6 days, measuring the dielectric breakdown voltage and partial discharge inception voltage of the samples. The partial discharge inception voltage of the modified mineral insulating oil was 1.16 times that of the unmodified mineral insulating oil after aging, and the breakdown voltage was up to 8 kV higher than that of the unmodified mineral insulating oil. These results show that the dielectric breakdown voltage and partial discharge inception voltage of mineral insulating oil can be effectively improved by modification with nano-TiO_2. The above studies illustrate that nano-TiO_2 has a positive effect on the breakdown voltage of the mineral insulating oil.

SiO_2 nanoparticles (denoted nano-SiO_2), as shown in Figure 8, are also often used as insulating nanoparticles to modify insulating oil. Muhammad and Li [73] added nano-SiO_2 to Karamay No. 25 mineral oil to give a volume fraction of nano-SiO_2 of 20%. Their measurement results showed that the AC breakdown voltage of the nano-SiO_2-modified insulating oil was higher than that of the unmodified mineral insulating oil, but the breakdown voltage decreased as the humidity of samples increased.

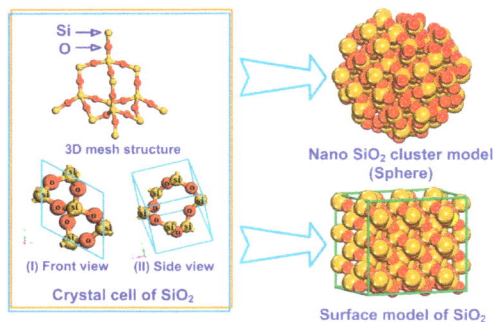

Figure 8. The molecular model of nano-SiO$_2$.

Ma et al. [74] added 1 wt % nano-SiO$_2$ to mineral insulating oil. The sample was aged of 35 days at 100 °C, and parameter testing was carried out every 7 days. They found that the addition of nano-SiO$_2$ increased the breakdown voltage of mineral insulating oil during the aging process. The breakdown voltages of mineral insulating oil modified with nano-SiO$_2$ and nano-TiO$_2$ are presented in Table 2.

Similarly, Al$_2$O$_3$ has been widely studied as an insulating nanomaterial. Ajay and Purbarun [75] added Al$_2$O$_3$ particles with a size of 25–125 nm to mineral insulating oil. The effects of the concentration, morphology, permittivity, and size of the nanoparticles on the breakdown characteristics of the modified oil were studied.

For nonconductive nano-Al$_2$O$_3$, the polarization of the dielectric nanoparticles will produce a potential trap under the influence of an external electric field, which slows down the fast-moving electrons and converts them to negatively charged nanoparticles. Moreover, the higher mobility of electrons will lead to the greater shielding effect of nanoparticles.

The performance of nano-Al$_2$O$_3$-modified insulating oil has been measured with different electrode materials [76]. It was found that the breakdown voltage of nano-Al$_2$O$_3$-modified mineral insulating oil was higher than that of the unmodified mineral insulating oil. The maximum breakdown voltage was observed when concentration of nano-Al$_2$O$_3$ was 20 mg/L. The molecular structure of nano-Al$_2$O$_3$ is depicted in Figure 9.

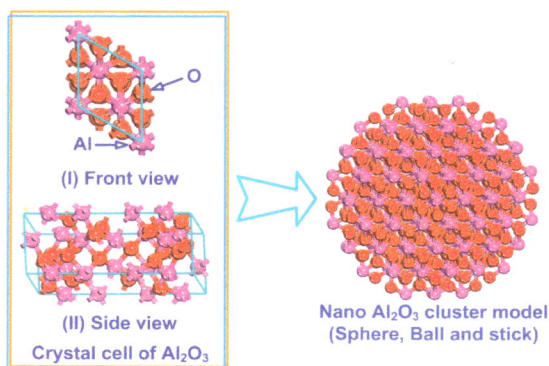

Figure 9. Molecular structure of nano-Al$_2$O$_3$.

Table 2. Breakdown voltages of mineral insulating oil modified with nano-TiO$_2$ and nano-SiO$_2$.

Nano-Particles	Year, Author	Manufacture Factory, Modified Object	Measurement Standard	Optimal Nanoparticles Concentration	Breakdown Voltage (kV)
TiO$_2$	2010, Du Y F, et al. [69]	Mineral oil	AC, IEC 60156, brass spherical electrodes, 2.5 mm gap	0.01 mg/L	57.8
	2011, Du Y F, et al. [70]	Mineral oil	AC, IEC60156, brass spherical electrodes, 2.5 mm gap	0.006 g/L	82.48
	2011, Du Y F, et al. [72]	Filtered mineral oil	AC, IEC60156, brass spherically-capped electrodes, 2 mm gap	0.075% (volume fractions)	80.9 (aged at 130 °C, 6 days)
	2016, Wang Q, et al. [71]	Petro China, No. 25 Kelamayi, filtered mineral oil	Impulse breakdown, IEC 60897, needle sphere electrode, 25 mm gap	10% *w/v*	About 78 (Positive impulse breakdown)
SiO$_2$	2016, Rafiq M, et al. [73]	Petro China, No. 25 Kelamayi, filtered mineral oil	AC, IEC 60897, brass spherical electrodes, 2 mm gap	20% (volume fractions)	About 76
	2016, Jun M, et al. [74]	Petro China, No. 25 Kelamayi, mineral oil	GB/T 507—2002	1 wt %	63.0 (110 °C, 0 day) 54.1 (110 °C, 14 days) 38.5 (110 °C, 35 days)

The average AC breakdown voltage of the Fe_3O_4 nanoparticle (nano-Fe_3O_4)-modified mineral insulating oil with an optimal nano-Fe_3O_4 concentration is 1.26 times that of the unmodified oil [77], and its average lightning breakdown voltage is 24% higher than that of the unmodified mineral insulating oil. The performance of the natural ester and nano-Fe_3O_4-modified mineral insulating oil was compared [78]. The experimental results showed that the nano-Fe_3O_4-modified mineral oil exhibited better performance than a mixture of natural esters and mineral oil from the aspects of dielectric properties and breakdown voltage. Therefore, nanofluids are considered as potential alternatives to conventional dielectric fluids. The breakdown voltage parameters of mineral insulating oil modified with nano-Fe_3O_4 and nano-Al_2O_3 are shown in Table 3. The molecular structure of nano-Fe_3O_4 particles is illustrated in Figure 10.

The above studies show that the addition of nanomaterials to mineral insulating oil can effectively increase its breakdown voltage. The mechanism behind this effect has been analyzed from the angle of microelectronics. Most researchers believe that the interfacial characteristics between the mineral insulating oil and nanoparticles play a dominant role in the space-charge transport during the breakdown process in nanofluids [73]. The oil–nanoparticle interface contains a large number of electronic traps, which repeatedly capture and release electrons. Such trapping lowers the electron mobility and energy transformation, and hinders the further development of the streamer in the electron capture and release process. In addition, both positive and negative pulse penetration tests have revealed that the addition of nanoparticles increases the positive and negative pulse penetration voltage of mineral insulating oil. The nanoparticles can absorb impurities in the mineral insulating oil, which suppresses the bridging effect and increases the breakdown voltage of the mineral insulating oil.

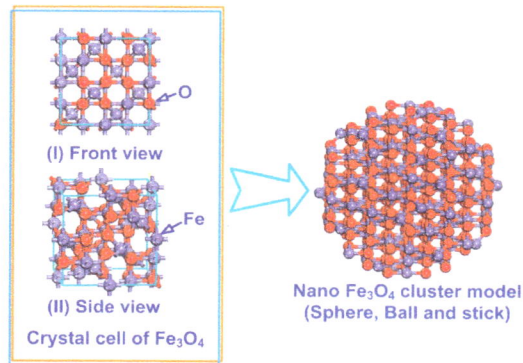

Figure 10. Molecular structure of nano-Fe_3O_4.

Table 3. Breakdown voltages of nano-Al_2O_3-modified and nano-Fe_3O_4-modified mineral insulating oils.

Nano-Particles	Year, Author	Manufacture Factory, Modified Object	Measurement Standard	Optimal Nanoparticles Concentration	Breakdown Voltage (kV)
Al_2O_3	2016, Katiyar A, et al. [75]	Mineral oil	AC, ASTM D-877, hemispherical electrodes, 5 mm gap	0.25 wt % (r = 23 nm)	About 68
	2016, Wang Q, et al. [71]	Petro China, No. 25 Kelamayi, filtered mineral oil	Impulse breakdown, IEC 60897, needle sphere electrode, 25 mm gap	20% *w/v*	About 85 (Positive impulse breakdown)
	2017, Qing Y, et al. [76]	Petro China, No. 25 Kelamayi, filtered oil	Impulse breakdown, brass electrodes, 1 mm gap	20 mg/L	About 40 (Impulse breakdown)
	2012, Zhou J Q, et al. [77]	Petro China, No. 25 Karamay, filtered mineral oil	AC, IEC 60156, brass spherical electrodes, 2.5 mm gap	1%	83.2
Fe_3O_4	2016, Wang Q, et al. [71]	Petro China, No. 25 Kelamayi, filtered mineral oil	Impulse breakdown, IEC 60897, needle sphere electrode, 25 mm gap	10% *w/v*	About 82 (Positive impulse breakdown)
	2016, Peppas G D, et al. [78]	Public Power Corporation of Greece, Shell Diala S2 ZU-I Filtered mineral oil	AC, IEC 60156, brass steel Rogowski electrodes, 2.5 mm gap	0.008%	About 77.7

3.1.2. Effect of Nanoparticles on the Dielectric Properties of Mineral Insulating Oil

In addition to breakdown voltage, the dielectric properties are also important parameters that reflect the electrical properties of mineral insulating oil. A study of mineral insulating oil modified with the ceramic nanomaterials zirconia (ZrO_2) and TiO_2 showed that the dielectric dissipation factor of the modified mineral insulating oil was lower than that of the unmodified mineral insulating oil [79]. Compared with the dielectric dissipation factor of nano-ZrO_2-modified mineral insulating oil, the nano-TiO_2-modified mineral insulating oil was higher. This may be because of the higher relative permittivity of nano-TiO_2 particles than that of nano-ZrO_2 particles. The kinematic viscosities of the nano-TiO_2 fluids were higher than those of nano-ZrO_2 because the particle size of the nano-TiO_2 fillers are larger than that of nano-ZrO_2, so the fluid flow can be prevented more effectively. The molecular structure of nano-ZrO_2 is depicted in Figure 11.

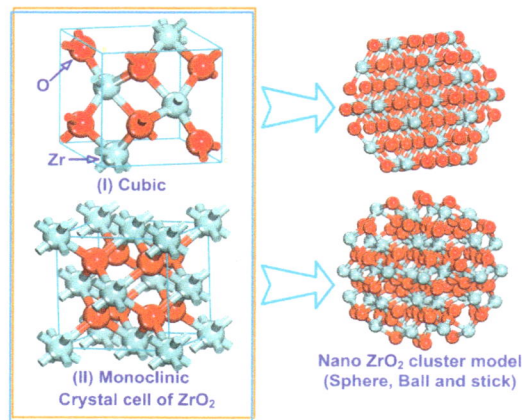

Figure 11. The molecular model of nano-ZrO_2.

A study of nano-TiO_2-modified DB-No 25 mineral insulating oil showed that a new low-frequency component (from 0.1 to 1 MHz) appeared in the dielectric frequency response of the nano-TiO_2-modified oil-paper insulating materials at different temperature and moisture contents [80].

In addition to nano-TiO_2, the improvement of the dielectric properties of mineral insulating oil induced by adding nano-SiO_2 has been investigated [81]. For these nanofluids, the dielectric withstand properties in a quasi-uniform field were enhanced when the nano-SiO_2 concentration was kept at about 0.2 g/L.

Ajay and Purbarun [75] modified mineral insulating oil with nano-Al_2O_3 with different particle sizes, and then measured the electrical parameters of the samples. The results showed that the dielectric properties increased by 69% when the content of nano-Al_2O_3 with a particle size of 23 nm was 0.25 wt %.

The above experiments reveal that the smaller the radius of the nanoparticles, the more obvious the improvement of the electrical properties of the modified mineral insulating oil. In addition, the electrical properties of modified mineral insulating oil tend to improve with increasing nanoparticle concentration.

3.1.3. Effect of Nanoparticles on the Thermal Stability of Mineral Insulating Oil

The internal temperature of a transformer during long-term operation is high. Therefore, mineral insulating oil needs to possess high thermal stability. Mineral insulating oil has been modified through the addition of boron nitride nanoparticles (nano-BN), as shown in Figure 12, to raise its thermal stability [82]. It was found that the thermal stability of the mineral oil modified with nano-BN was higher than that of the unmodified mineral insulating oil. When the nano-BN content was 0.1 wt %, the thermal conductivity of the modified oil increased continuously with rising temperature, and the increase was more than 70% when the temperature reached 27 °C.

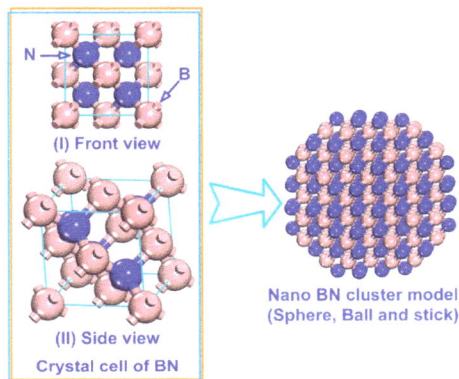

Figure 12. The molecular model of nano-BN.

Modification of mineral insulating oil with nanodiamond (ND) and subsequent measurements revealed that the thermal conductivity of the mineral insulating oil modified with a ND mass fraction of 0.12% was 14.5% higher than that of the unmodified insulating oil [83]. The change in the viscosity of the base fluid with a ND loading of up to 1% was quite small. It is noteworthy that greater enhancements of the thermal conductivity can be achieved by designing the ND covalent surface modifications, and optimizing the ND/base fluid solvation.

Cadena-dela et al. [84] studied the thermal properties of mineral oil-based nanofluids containing dispersed aluminum nitride nanoparticles (nano-AlN) and nano-TiO_2. Their results indicated that the thermal transfer coefficient of mineral insulating oil can be improved and the internal heat of the transformer can be easily dissipated through the addition of nanoparticles. When the of nano-TiO_2 content was 0.01 wt %, the kinematic viscosity was the lowest (15.80 m^2/s at 24 °C). A nano-AlN content of was 0.01 wt % gave the lowest kinematic viscosity of 15.82 m^2/s at 24 °C. At 40 °C, the same nano-TiO_2 and nano-AlN-modified samples displayed lowest kinematic viscosities of 7.21 and 7.32 m^2/s, respectively. In general, the viscosity of mineral insulating oil modified with nano-TiO_2 was lower than that with nano-AlN, and nano-TiO_2 also improved the thermal stability of the mineral insulating oil more than nano-AlN.

When Lv et al. [85] added nano-TiO_2 to mineral insulating oil, they found that the diameter of impurities was decreased, and the sample fluidity under a low electric field was greatly improved. The charged nanoparticles became the major transmission factors and slowly floated with increasing electrical stress because the mineral insulating molecules provided high viscous resistance.

The above analysis illustrates that the modification of mineral insulating oil with nanomaterials can obviously enhance the three properties of power frequency breakdown, partial discharge voltage, and positive impulse breakdown voltage. The strengthening the negative-polarity impulse breakdown voltage is affected by the surface modification method of nanoparticles, original oil samples, and testing method. Analysis has revealed that the addition of nanoparticles will decrease the resistivity of mineral insulating oil and increase its dielectric loss angle [86]. Typical dielectric properties of mineral insulating oil and nanofluids are listed in Table 4.

Table 4. Typical dielectric properties of mineral insulating oil and nanofluids.

Oils	Resistivity ($\Omega \cdot$m)	tan δ	Relative Permittivity (ε_r)
Pure oil	1.41×10^{12}	0.008	2.19
Fe_3O_4 nanofluids	2.05×10^{10}	0.360	2.35
TiO_2 nanofluids	2.53×10^{10}	0.488	2.90
Al_2O_3 nanofluids	2.56×10^{11}	0.046	2.27

3.2. Modification of Non-Nanoparticles

The limited availability of petroleum resources and increasingly serious environmental problems drive the search for alternatives to mineral insulating oil. Natural esters are considered an attractive green alternative to mineral insulating oil [1,87], and mixtures composed of natural esters and mineral insulating oil have become a research interest. In 2002, Fofana et al. [21] studied the electrical and physicochemical properties of mixtures of natural esters and mineral insulating oil with different proportions. They found that when the natural ester content was less than 20%, all the electrical and physical properties of the mixtures were better than those of the traditional transformer mineral insulating oil. When the ester content exceeded 50%, the density and viscosity exceeded the standard limits. During determination of the mass of mineral insulating oil, the density is generally not important; however, it does become important at lower temperature. Addition of natural esters to mineral insulating oil helps to suppress gasification under local thermal stress. In 2009, Liao et al. [30] found that natural esters display high water saturation, and can help to restrain oxidation reactions of the mixed oil when added to mineral insulating oil. However, the viscosity of natural esters is high, so an excessive content of natural esters will enhance the viscosity of the mixed oil. This conclusion is consistent with the results of Fofana et al. [21], who also researched the electrical, physicochemical, and degradation properties of mixtures of mineral insulating oil with natural esters (olive oil) with different components. They found that hydrolysis of the natural esters increased the acid content of the mineral insulating oil and decreased the breakdown voltage of the mixed oil. The oxidation stability of mineral insulating oil can be effectively improved by using an appropriate proportion of natural esters to mineral insulating oil and adding an antioxidant.

Test results have indicated that the biodegradability of mixed insulating oils is higher than that of mineral insulating oil and more environmentally benign. For example, Yusnida et al. [88] added palm oil to mineral insulating oil and researched the breakdown characteristics of the mixtures with different contents of palm oil. It was found that when the content of palm oil was less than 20%, the breakdown strength of the mixed oil decreased with increasing palm oil content. When the palm oil content was more than 20%, the breakdown strength of mixed oil increased with the palm oil content. When the content of palm oil was 80%, the maximum breakdown strength of 87 kV was observed. Moreover, the kinematic viscosity of the mixed oil at 40 °C rose with increasing palm oil content.

Rapp et al. [89] found that natural esters have a high affinity for water, and will allow more water to be transferred from the cellulose paper to the natural ester liquid. At the same time, the natural esters will consume the moisture in the fluid through hydrolysis and allow the moisture to reach a steady dynamic equilibrium between the cellulose paper and esters. However, free fatty acids are produced as secondary reactants of ester exchange reactions during the hydrolysis, which can change the structure of cellulose and degrade the electrical properties of the cellulose paper. In 2010, Liao et al. [90] found that the anti-aging ability of mixed oil-paper insulation with natural esters added to the mineral insulating oil was superior to that of mineral oil-paper insulation. The reason for this is that the natural esters in the mixed oil-paper insulation form stable hydrogen bonds with moisture and acid molecules, and weaken the synergistic hazardous effect of moisture and acid on the insulating paper. At the same time, the esterification of hydroxyl and fatty acids in insulating paper cellulose can inhibit the aging of the insulation paper [91]. In addition, the thermal cracking rate of the mixed oil-paper insulation was lower than that of mineral insulating oil-paper insulation.

In 2011, Liao et al. [92] pointed out that the main reason for the considerable inhibitory effect of mixed oil on the aging rate of oil-paper insulation is that the aged insulation paper generates new ester groups. These ester groups inhibit insulating paper thermal aging. The content of aldehyde groups generated during aging of mixed oil-paper insulation is lower than that generated in the mineral oil-paper insulation. It was found that the mixed oil can restrain its own oxidation and the degradation rate of the insulating paper, and also improve the thermal stability of mineral insulating oil-paper insulation systems [93]. Studies on the thermal aging of mixed oils with natural esters have revealed that the water and acid values of the mixed oil-paper insulation are higher than those of the mineral insulating oil-paper insulation, and the mixed oil-paper insulation has a higher breakdown voltage than that of the mineral oil-paper insulation [94].

In summary, natural esters possess the advantages of high flash point and ignition point, good electrical properties, high biodegradability, and high abundance and have thus been widely used as insulating materials. However, the kinematic viscosity of most natural ester fluids is high, which will adversely affect the heat emission of a transformer when used as insulating oil. In addition, the acid content of natural esters after aging is higher than that of traditional mineral insulating oil, affecting the performance of mineral insulating oil-paper insulation. When natural esters are added to mineral insulating oil, the natural esters and the aged insulation paper undergo chemical reactions to form ester groups, and restrain the aging of the oil-paper insulation system and prolongs the service life of the mineral insulating oil-paper insulation. Therefore, the appropriate proportions of mineral insulating oil and natural esters can realize mixed oils that possess the complementary advantages of both materials.

4. Application of Computer Simulation Technology

In 1901, Gibbs introduced the concept of an ensemble [95], which led to a big step forward in molecular calculations based on statistical mechanics. In 1947, the creation of electronic computers made large-scale scientific calculation possible. Since then, computational molecular science has developed rapidly and is now widely used in many fields, like chemical engineering, material engineering, pharmaceutical engineering, and life sciences, because of its great potential and strength [96]. Computer-based molecular simulations are usually conducted by one of two methods: the Monte Carlo method (MC) [97] or the molecular dynamics (MD) method [98], although quantum mechanics and molecular mechanics methods also exist [99–103]. Although there has been almost no change of the central algorithm of molecular simulation since the 1950s, the constant update of optimization algorithms means that the efficiency of molecular simulation is continuously improving.

The dynamic properties simulated by MD calculations have been widely used in materials and life sciences. The interaction modes and energies of molecules, atoms, or ions; energy components like bond energy and angle energy; gyration radius; tension; pressure; volume; and cell parameters can be calculated for multimolecular, macromolecular, and solution systems. Through the analysis of MD calculation results, the radial distribution function (RDF), mean square displacement (MSD) and fluctuation analysis results can be obtained. As a result, the stacking properties, orientation, compression properties, and phase transition properties of materials and systems can be acquired.

In recent years, the concept of high-voltage engineering calculations has been developed. In this research category, the quantitative analytical techniques used in many interdisciplinary subjects including electrical science, quantum mechanics, material physics, and computational chemistry are integrated. Numerical simulation and analysis can be carried out to investigate scientific and technical problems in high voltage engineering using electronic computers and the discrete numerical method, forming new interdisciplinary branches [104].

Molecular simulation technology can provide basic methodology and quantitative analysis methods to reveal the microphysical and chemical properties of electrical insulation, resolving the insulation degradation and destruction mechanism of power equipment. According to the development status of high-voltage and insulation technology, the computer simulation methods of multiphysics numerical simulation and electromagnetic transient analysis are combined to achieve effective theoretical support for high-voltage engineering practice. In recent years, great success has been obtained in the study of the performance of mineral oil-paper insulation systems at molecular and atomic levels using molecular simulation.

4.1. Molecular Simulation of Water and Acid in Oil Paper Insulation System

In 2007, Lu [105] studied the effects of water and acid on mineral oil-paper insulation systems by simulations using the Condensed-phase Optimized Molecular Potential for Atomistic Simulation Studies (COMPASS) force field. In 2009, Liang [29] researched the anti-aging performance of transformer oil and the effect of the insulating paper on thermal aging. It was found that the ester exchange reaction between the polymer acid and cellulose in the mixed oil and hydrogen bonds generated between the ketonic oxygen atoms of the esters and water molecules provided joint anti-aging ability. The degradation rate of the insulating paper in the mixed oil was obviously slower than that of the insulating paper in mineral insulating oil. The biological degradability of environmentally friendly anti-aging transformer oil and thermal aging properties of oil-paper insulation have been researched [106]. The binding of environmentally friendly anti-aging oil with water molecules is stronger than that of mineral oil, the aging of mixed oil is obviously slower than that of mineral insulating oil, and the polymerization degree of mixed oil-impregnated insulating paper is obviously higher than that mineral oil-impregnated insulating paper.

In 2010, Liao et al. [90] investigated the effects of water and acid on thermal aging of mineral insulating oil-cellulose insulating paper and mixed oil–cellulose insulating paper. Their results showed that the anti-aging ability of the mixed oil–insulation paper was superior to that of the mineral insulating oil-paper; the natural esters in the mixed oil-paper system containing water and acid can form stable hydrogen bonds, which suppress the synergistic hazardous effect of moisture and acid on the insulating paper. In the mixed oil-paper insulation system, the esterification of hydroxyl and fatty acids in the insulating paper cellulose inhibited the aging of the insulation paper. The group found that the molecular conformations of oil on different surfaces of the cellulose were varied through analysis of the mutual interactions between the mineral insulating oil-paper insulation materials. The conformation of oil molecules and cellulose molecules at the interface of an amorphous region is shown in Figure 13. The diffusion of moisture in the paper was obstructed because of the change of mineral insulating oil density at the surface of the cellulose crystal; this phenomenon will accelerate the diffusion of moisture in the paper [107]. Then same group then researched the effect of initial moisture content on the thermal properties of mineral oil-paper insulation [108]. Their results showed that the diffusion capacity of furfural and small acid molecules obviously decreased with increasing moisture content. This is because moisture, furfural, and small acid molecules generate stable hydrogen bonds. At the same time, the interaction energy is changed because of the polarity of these species.

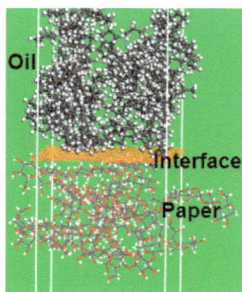

Figure 13. Simulated conformation of oil and cellulose molecules in the cellulose amorphous region at an oil-paper interface.

4.2. Molecular Simulation of Thermal Cracking

The pyrolysis of mineral insulating oil is one of its main types of degradation when internal insulation defects occur in transformers. Molecular simulation is an important tool to study microscopic mechanisms. The pyrolysis mechanism of mineral oil-paper insulation systems has been studied by molecular simulation in recent years. The results provide potential guiding value to further understand the cracking process of oil-paper insulation systems and evaluate the insulation life of the transformer after overheating defects form.

In order to research the generative mechanisms associated with the pyrolysis of triglycerides, Zhang et al. [109] carried out 500 ps MD simulations on a tripalmitin model by using a reactive force field (ReaxFF) at 1500 and 2000 K. In 2016, Zhang [110] used a molecular simulation method based on ReaxFF to study the microcosmic mechanism of the pyrolysis process of mineral oil-paper insulation, and clarified the initial cracking mechanism of the insulating paper and the formation mechanism of the main products through the mutual verification of theories and experiments; meanwhile, the micro dynamic mechanism of the thermal cracking of insulating paper was studied at the atomic level. In 2016, Lin et al. [111] studied the reaction mechanism of thermal cracking of transformer oil through molecular simulation, established a simulation model of mineral insulating oil molecules at different temperatures, and studied the generation rules of gas molecules in oil during transformer pyrolysis. The micro cracking mechanism of the three typical components of alkanes, cycloalkanes, and aromatic hydrocarbons in transformer oil was studied using ReaxFF to determine the relationship between temperature and the pyrolysis process. The proposed kinetic mechanism of thermal decomposition of mineral insulating oil at the atomic level was consistent with the pyrolysis results obtained from experiments.

In 2017, Wang et al. [4] tried to reveal the dynamic reaction mechanism at an atomic level using the ReaxFF to simulate reactive MD of mineral insulating oil pyrolysis at high temperature and the influence of acid in the mineral insulating oil on pyrolysis. It was found that as the temperature rises, the cracking rate of all three reactants considered in the paper will speed up markedly and the pyrolysis products are mainly small molecules and radicals. The reaction paths revealed that under acidic conditions, H atom is mainly produced by the reaction between hydroxyl H in formic acid and H in hydrocarbons, which is found to be the reason for the acceleration of pyrolysis of mineral insulating oil by formic acid.

4.3. Molecular Simulation of Small Molecular Diffusion

The diffusion behavior of soluble materials in mineral insulating oil (including moisture and gas molecules) and its effect on the properties of mineral oil-paper insulation has been investigated. In particular, the diffusion behavior of small molecules in oil-paper insulation has been considered [105,112,113]. The free volume theory has been proposed to explain the diffusion and mass transfer phenomenon for gas in mineral insulating oil on the basis of the diffusion of gas produced by the aging of mineral oil-paper insulation systems [105]. The free volume theory is of great important to understand the diffusion behavior of small gas molecules in mineral insulating oil. In other work, a microcosmic model of insulating paper and oil paper was created based on the molecular simulation method [112]. The movement trajectory and diffusion coefficient of moisture in this model at different temperatures was calculated through MD. The relationship between the diffusion coefficient, free volume in the model, and movement trajectory of water molecules was examined. The diffusion coefficient of the water molecules was obtained experimentally, which verified the calculation results. The results obtained by the molecular simulation were 84–222% of the values of experimental data. The microscopic mechanism of gas molecular diffusion has been analyzed through diffusion coefficients, displacement characteristics, free volumes, and interactions [114]. The differences in the diffusion characteristics of different gas molecules were discussed, and the factors that influence gas molecules diffusion were compared; these results are summarized in Tables 5 and 6 (D is the diffusion coefficient, a is the slope of the curve, R^2 is the goodness of fit). Studies show that the diffusion coefficient of gas molecules in cellulose is lower than that in oil by one order of magnitude, and the diffusion coefficients of the gas molecules in two insulating media are of different orders. The free volume is the main factor that influences the gas diffusion behavior in oils, whereas the intermolecular interaction is the main factor that influences the diffusion behavior of celluloses.

Table 5. Diffusion coefficients of gas molecules in oil ($\text{Å}^2/\text{s}$) [114].

Parameter	H_2	CH_4	C_2H_4	C_2H_2	C_2H_6	CO	CO_2
a	10.886	1.3875	2.7873	2.2322	2.0577	3.4631	1.6641
R^2	0.9898	0.9640	0.9976	0.9848	0.9933	0.9729	0.9822
D	1.8143	0.4645	0.2773	0.3720	0.2313	0.5772	0.3429

Table 6. Diffusion coefficients of gas molecules in cellulose ($\text{Å}^2/\text{s}$) [114].

Parameter	H_2	CH_4	C_2H_4	C_2H_2	C_2H_6	CO	CO_2
a	0.3491	0.1893	0.0523	0.1297	0.1618	0.3158	0.2655
R^2	0.8682	0.9230	0.5049	0.7766	0.8976	0.8628	0.9601
D	0.0582	0.0316	0.0087	0.0216	0.0270	0.0526	0.0443

4.4. Molecular Simulation of Nanoparticle Modification

In recent years, increasing attention has been paid to the use of nanomaterials in the oil-paper insulation of transformers. Many studies have been conducted on nanoparticle modification as a method to improve the performance of mineral oil-paper insulating systems, including their thermal, mechanical, and insulating properties, and good results have been achieved.

In 2010, Christopher, A et al. [113] performed electrostatic field simulations to study the effect of barium strontium titanate nanoparticle suspensions on the electric fields within synthetic insulating oil. The simulations confirmed that, by adding high dielectric constant nanoparticles, the electric field of insulating oil can be changed dramatically through charge polarization. The nanoparticles are able to generate large electron extraction fields on the cathode surface and form paths of higher electric field across the gap, which may help to minimize streamer propagation jitter.

Dai et al. [115] studied the dispersion stability of nanoparticles in mineral insulating oil through MD simulation. Their results showed that nano-Al_2O_3 with a diameter of 18 nm forms a stable dispersion in mineral insulating oil. In 2015, Shi et al. [116] studied the effect of nanoparticles on the properties of mineral oil-paper composite insulation systems by conducting experiments and simulation studies. They also compared the operating impulse breakdown voltages of the mineral oil-paper composite insulation system with and without nanoparticles. The group developed a model to describe fluid injection in the oil-paper system, and calculated the electric field in the fluid injection channel along the surface, space charge, and distribution of surface charge along the surface of the insulating paper. It was found that the voltage withstand properties of the modified insulation paper were improved by about 10% compared with those of the unmodified insulation paper. The surface charge density decreased from 0.020 to 0.016 C/m^2 upon nanoparticle modification, and the nanoparticles restricted the development of fluid injection in mineral insulating oil and fluid injection along the surface, as well as improving the insulation properties of the insulating paper. In 2015, Adil et al. [117] investigated the system's rheological properties and diffusion coefficient, and studied the nanocluster's disperse and stability. The calculated viscosity of the CuO-alkane system was 1.613 mPa s at 303 K.

Zhou et al. [118] studied the heat transfer characteristics of nanomodified mineral insulating oil. To do this, they prepared tnano-SiO_2-modified transformer oil with different nano-SiO_2 concentrations, and compared their thermal conductivity values. The results showed that the thermal conductivity of modified mineral insulating oil will gradually increase with rising nano-SiO_2 concentration. Computational models of different kinds of nanoclusters (nano-SiO_2, nano-Al_2O_3, and nano-ZnO) in mineral insulating oil fluidic systems have been constructed at the atomic molecular level to investigate the effect of temperature on the system viscosity [119]. Simulation results showed that the viscosities of the pure mineral insulating oil and nanofluids decrease as the increase of temperature, and the relationship of viscosity with temperature does not change after adding nanoparticles. The viscosities of nanofluids are higher than that of pure oil, and nano-SiO_2 is the most suitable nanoparticle additive to minimize the increase of viscosity. The viscosity of both pure and nanomodified mineral insulating oil decreases with increasing temperature [120]. The molecular structure of nano-ZnO is shown in Figure 14.

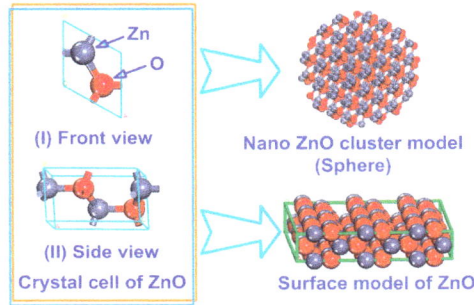

Figure 14. Molecular structure of nano-ZnO.

Moreover, the structural and dynamical properties at the water/oil interface in various systems have been investigated by the dissipative particle dynamics simulations [121]. For all nanoparticles used in literature, as expected, a transition from the liquid-like state to the solid-like state with the surface density increases was observed. However, at the water/oil interface, different nanoparticles have different contact angles, which made the results for systems containing mixtures of nanoparticles more complex.

Shi et al. [122] performed MD simulations to study the microscopic behaviors of anionic, nonionic, and zwitterion at an oil/water interface. Their results showed that these four kinds of surfactants can form a stable monolayer at an oil/water interface. Possover et al. [123] pointed out that because nanofluids have excellent thermal performance owing to their high thermal conductivity, they may show promise as nanofluid coolants that are electrically insulating. By using equilibrium MD simulations followed by the application of the Green-Kubo autocorrelation function, they investigated the thermal conductivity of a BN suspension.

Riku et al. [124] performed a number of MD simulations on nonionic nanoparticle/surfactant systems. Analysis of the results allowed the dispersive interactions of the nanoparticles and surfactants to be related to their physical behaviors at the oil/water interfaces.

Our team studied the effect of nano-SiO_2 on water diffusion and its mechanism in naphthenic-based mineral insulating oil using the MD method. We established a model of mineral insulating oil modified with nano-SiO_2 and different moisture contents (denoted as the modified model in Figure 15a and a model of mineral oil without nano-SiO_2 (denoted as the unmodified model, as shown in Figure 15b, and performed MD simulations to calculate the microparameters of all the models after full relaxation. Figure 15c,d present the statistical results for the Connolly surfaces of the modified and unmodified models, respectively. Preliminary findings indicate that in the case of normal operation of a transformer, addition of nano-SiO_2 will slow the diffusion of water molecules in mineral insulating oil through the adsorption of water molecules in mineral oil, therefore decreasing the probability that water molecules form a small bridge. This simulation revealed why the breakdown voltage of the mineral insulating oil is improved through the addition of nano-SiO$_2$ in macroscopic experiments.

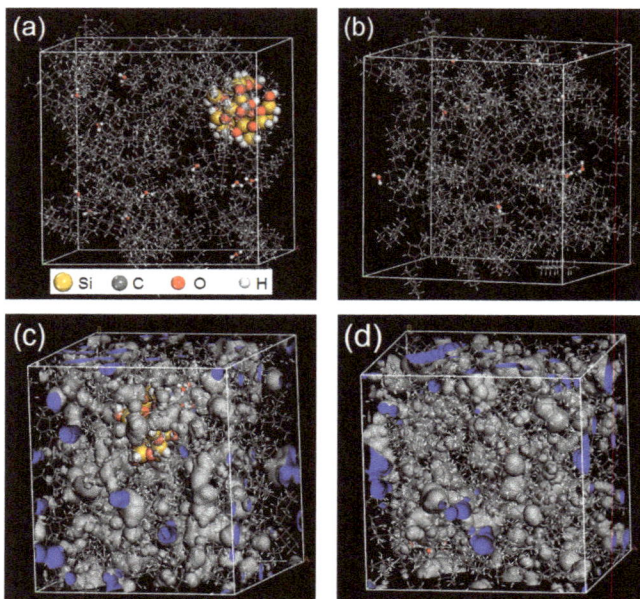

Figure 15. Effect of water on the properties of mineral insulating oil: (**a**) Modified model; (**b**) Unmodified model; (**c**) Statistical results for the Connolly surface of the model in (**a**); (**d**) Statistical results for the Connolly surface of the model in (**b**).

The above examples illustrate that molecular simulation technology plays an important role in the research of mineral insulating oil for transformers. On one hand, molecular simulation technology can explain the appearance of macroscale phenomena, such as the influence of nanoparticles on the viscosity of mineral insulating oil and the temperature dependence of water diffusion in mineral insulating oil. On the other hand, molecular simulation technology provides powerful guidance on how to further improve the performance of mineral insulating oil, such as the type of nanoparticle added, determination of the amount added, and the components in the mixture.

5. Conclusions and Prospects

As research methods for traditional mineral oils have gradually matured, the composition, partial aging, and cracking mechanism of mineral insulating oil were initially learned, and some factors that affect its performance were then studied. On this basis, ways to improve the performance of mineral insulating oil have been proposed. Even so, there are still many problems that are worth exploring, which are summarized as follows:

(1) With the rise of the operating voltage of power grids, the requirements for the insulation, mechanical, and heat resistance properties of transformers are also gradually increasing. Therefore, it has become important to develop mineral oils with better performance.

(2) Because mineral insulating oil is a non-renewable resource, it will face shortage and exhaustion issues. At the same time, in the face of growing environmental problems, the use of mineral insulating oil will be further restricted because of its poor biodegradability and environmental risk. Recent studies have shown that vegetable oils are completely biodegradable and

pollution-free [125,126]. Treated plant insulating oil, such as sunflower oil, olive oil, and rapeseed oil, has the advantages of high flash point and large dielectric permittivity, so it is a good substitute for mineral insulating oil [127–129]. However, most vegetable oils possess high viscosity, pour point, and acid content after aging compared with those of mineral oil. Thus, electrical and mechanical equipment will be affected to a certain extent during long-term operation in vegetable oil at high temperature and high pressure [130–132].

(3) The micromechanics of the discharge process of nanomodified mineral insulating oil is worth studying. For example, the creeping process along the surface of paperboard immersed in mineral insulating oil should be analyzed to reveal the effect of nanoparticles on the microstructure of the mineral insulating oil-paper interface and explore the effect of nanoparticles on streamer development at the mineral insulating oil-paper interface.

(4) The macroscopic adaptability of nanomodified mineral insulating oil needs to be investigated further. The addition of different nanoparticles influences the improvement of the thermal conductivity of the mineral insulating oil and its overall heat transfer efficiency. Because the mineral oil is used as an insulating medium for the long-term operation of transformers, its insulating, electrical, anti-aging, moisture, and stability (degree of nanoparticle agglomeration) characteristics are all important factors that influence its performance and application.

(5) The mechanism of nanomodified mineral insulating oil and insulating paper during operation needs to be studied further. The internal insulation of a transformer is mainly a hybrid insulation system consisting of insulating paper and mineral insulating oil. However, there has been no in-depth research on the influence laws and mechanism of nanoparticles on oil-paper hybrid insulation systems. In addition, little research has been reported on the electrical properties and stability of nanomodified oil-paper systems under the conditions of long-term electric/thermal aging and high moisture content. The dielectric constant of oil-paper systems is changed by the addition of nanoparticles, which greatly influences the distribution of electric field in the transformer. However, the mechanism of this effect is currently unclear, so there is a need to focus on the effect of nanoparticles on oil-paper hybrid insulation systems in future research.

(6) Multidisciplinary integration is a current trend to develop basic theory in mineral insulating oil research. At the macroeconomic level, research on mineral insulating oil involves the basic subjects of classical physics, such as classical mechanics, photonics, and electromagnetism, and at the micro level, the research involves molecular chemistry, surface science, quantum mechanics, and the theory of relativity. The macroscopic behavior and microscopic mechanism of mineral insulating oil must be integrated to fully understand its characteristics.

(7) Molecular simulation technology will still play an important role in investigating the aging, cracking, and other microscale behaviors of mineral insulating oil. The properties of mineral insulating oil are affected by the complex internal environment of the transformer. Current molecular simulations still require the transformer to have a specific state, so the operating environment of the transformer cannot be completely simulated. Therefore, molecular simulation technology needs to be further developed to better reflect the real operation environment of mineral insulating oil in combination with multifactorial simulation. With the further development of computer technology, especially advances in high-performance calculation, it should be possible to build bigger and more optimized models. Faster operation has also become a trend in the development of molecular simulation technology.

Acknowledgments: The authors wish to thank the guest editor Issouf Fofana for the kind invitation to present this review article in the Special Issue "Engineering Dielectric Liquid Applications". Besides, the authors wish to thank the National Key R&D Program of China (Grant No. 2017YFB0902700, 2017YBF0902702) for their financial support.

Author Contributions: All authors collected, organized and analyzed the references; Xiaobo Wang and Chao Tang wrote the paper, while other authors offered their modification suggestions for the manuscript.

Conflicts of Interest: The authors declare no conflict of interest.

References

1. Oommen, T.V. Vegetable oils for liquid-filled transformers. *IEEE Electr. Insul. Mag.* **2002**, *18*, 6–11. [CrossRef]
2. Fofana, I. 50 years in the development of insulating liquids. *IEEE Electr. Insul. Mag.* **2013**, *29*, 13–25. [CrossRef]
3. Liu, F.L.; Xu, W. Characteristic Comparison between paraffine-Base and naphthene-Base transformer oils. *Transformer* **2004**, *41*, 6–8.
4. Wang, X.L.; Li, Q.M.; Zhang, Y.; Yang, R.; Gao, S.G. Simulation of reactive molecular dynamics of transformer oil pyrolysis at high temperature and the influence mechanism of acid in oil. *High Volt. Eng.* **2017**, *43*, 247–255.
5. Ma, S.J.; Zhang, L.J.; Yang, J.J. Application of naphthenic base oil to transformer. *Transformer* **2005**, *42*, 28–31.
6. Wang, Y.G. Low temperature flow characteristics of naphthenic-base and paraffinic-base oils in transformers and circuit breakers. *North China Electr. Power* **1988**, *1*, 37–46.
7. Shen, B.W. Discussion on naphthenic base oil. *Pet. Prod. Appl. Res.* **2007**, *25*, 46–48.
8. Chen, Y. Development of transformer oil products and standards at home and abroad. *Pet. Prod. Appl. Res.* **2009**, *26*, 16–24.
9. Zheng, P.Y.; Qin, H.N. Characteristics, processing and application of naphthenic base oil. *Mod. Bus. Trade Ind.* **2008**, *20*, 267–269.
10. Mulhall, V.R. The significance of the density of transformer oils. *IEEE Trans. Electr. Insul.* **1980**, *15*, 498–499. [CrossRef]
11. Gunnar, S. Polycyclic aromatic hydrocarbons—Key to the characteristics of transformer oil. *Transformer* **1996**, *7*, 6–7.
12. Chen, W.Y.; Tang, H.J.; Ding, T.T.; Zhou, X.Z. The characteristics of naphthenic crude oil and the analysis of supply and demand of naphthenic base oil market. *Pet. Prod. Appl. Res.* **2013**, *31*, 47–51.
13. Ekonomou, L.; Skafidas, P.D.; Oikonomou, D.S. Transformer oil's service life identification using neural networks. In Proceedings of the 8th WSEAS International Conference on Electric Power Systems, High Voltages, Electric Machines (POWER' 08), Venice, Italy, 21–23 November 2008; pp. 222–226.
14. Dervos, C.T.; Vassiliou, P.; Skafidas, P.; Paraskevas, C. Service life estimation of transformer oil. In Proceedings of the International Conference on Protection and Restoration of the Enviroment VI, Skiathos, Greece, 4–5 July 2002; pp. 1239–1246.
15. Tokunaga, J.; Koide, H.; Mogami, K.; Hikosaka, T. Gas generation of cellulose insulation in palm fatty acid ester and mineral oil for life prediction marker in nitrogen-sealed transformers. *IEEE Trans. Dielectr. Electr. Insul.* **2017**, *24*, 420–427. [CrossRef]
16. Li, S.B.; Gao, G.Q.; Hu, G.C.; Gao, B.; Gao, T.S.; Wei, W.F.; Wu, G.N. Aging feature extraction of oil-impregnated insulating paper using image texture analysis. *IEEE Trans. Dielectr. Electr. Insul.* **2017**, *24*, 1636–1645. [CrossRef]
17. Mahanta, D.K.; Laskar, S. Electrical insulating liquid: A review. *J. Adv. Dielectr.* **2017**, *7*, 1730001. [CrossRef]
18. Emsley, A.M.; Stevens, G.C. Review of chemical indicators of degradation of cellulosic electrical paper insulation in oil-filled transformers. *IEE Proc. Sci. Meas. Technol.* **1994**, *141*, 324–334. [CrossRef]
19. Lundgaard, L.E.; Hansen, W.; Linhjell, D.; Painter, T.J. Aging of oil-Impregnated paper in power transformers. *IEEE Trans. Power Deliv.* **2004**, *19*, 230–239. [CrossRef]
20. Shorff, D.H.; Stannett, A.W. A review of paper aging in power transformers. *IEEE Proc. C* **1985**, *132*, 312–314. [CrossRef]
21. Fofana, I.; Wasserberg, V.; Borsi, H.; Gockenbach, E. Challenge of mixed insulating liquids for use in high-voltage transformers.1. Investigation of mixed liquids. *IEEE Electr. Insul. Mag.* **2002**, *18*, 18–31. [CrossRef]
22. Liu, Y.X. Water Accumulation in oil-paper insulation transformer and its influence to life time. *Transformer* **2004**, *41*, 8–13.
23. Chen, W.G.; Gan, D.G.; Liu, Q. On-line monitoring model based on neural network for moisture content in transformer oil. *High Volt. Eng.* **2007**, *33*, 73–78.
24. Yang, L.J.; Liao, R.J.; Sun, C.X.; Jian, L.I.; Liang, S.W. Using multivariate statistical method to recognize different aging stages of oil-paper. *Proc. CSEE* **2005**, *25*, 81–84.

25. Liao, R.J.; Sun, H.G.; Yin, J.G.; Gong, J.; Yang, L.J.; Zhang, Y.Y. Influence on the thermal aging rate and thermal aging characteristics caused by water content of oil-paper insulation. *Trans. China Electrotech. Soc.* **2012**, *27*, 34–42.

26. Zhou, L.J.; Tang, H.; Zhang, X.Q.; Wu, G.N. Disequilibrium moisture partition in oil-paper insulation. *Proc. CSEE* **2008**, *28*, 134–140.

27. Mcshane, C.P.; Gauger, G.A.; Luksich, J. Fire resistant natural ester dielectric fluid and novel insulation system for its use. In Proceedings of the IEEE Transmission and Distribution Conference, New Orleans, LA, USA, 11–16 April 1999; pp. 890–894.

28. Mcshane, C.P. Natural and synthetic ester dielectric fluids: Their relative environmental, fire safety, and electrical performance. In Proceedings of the IEEE Industrial and Commercial Power Systems Technical Conference, Sparks, NV, USA, 2–6 May 1999; pp. 1–8.

29. Liang, S.W. Study on Antioxidation Transformer Insulation Oil and Its Effects on Oil-Immersed Paper Thermal Aging. Ph.D. Thesis, Chongqing University, Chongqing, China, 2009.

30. Liao, R.J.; Liang, S.W.; Li, J.; Hao, J.; Yin, J.G. Study on the Physics and Chemistry Characteristics and Breakdown Voltage of a Mixed Insulation Oil Composed of Mineral Oil and Natural Easter. *Proc. CSEE* **2009**, *29*, 117–123.

31. Yin, J.G. Moisture Transfer Rule of Oil-Paper Insulation and Its Impact on the Ageing Characteristics during Thermal Ageing Process. Master's Thesis, Chongqing University, Chongqing, China, 2010.

32. Liao, R.J.; Zhang, S.; Yang, L.J.; Lin, B.; Hao, J. Comparative Analyses of Thermal Aging and Power Frequency Breakdown Characteristics between Natural Ester-paper Insulation and Mineral Oil-paper Insulation. *High Volt. Eng.* **2012**, *38*, 769–775.

33. Guo, P. The Therming Aging Properties of the New Mixture Oil-Pressboard Insulation and Its Application in the Distribution Transformer. Master's Thesis, Chongqing University, Chongqing, China, 2014.

34. *Insulating Liquids—Determination of the Breakdown Voltage at Power Frequency*; National Standard, GB/T 507-2002; National Energies Administrator: Beijing, China, 2002.

35. Darveniza, M.; Saha, T.K.; Hill, D.J.T.; Le, T.T. Investigations into effective methods for assessing the condition of insulation in aged power transformers. *IEEE Trans. Power Deliv.* **1998**, *13*, 1214–1223. [CrossRef]

36. Lundgaard, L.E.; Hansen, W.; Ingebrigtsen, S. Ageing of mineral oil impregnated cellulose by acid catalysis. *IEEE Trans. Power Deliv.* **2008**, *15*, 540–546. [CrossRef]

37. Xu, J.; Zhang, X.M.; Han, M.; Zhang, H. Study of the experiment of breakdown of electric insulation oil. *J. Chongqing Technol. Bus. Univ.* **2008**, *25*, 71–73.

38. Koch, M.; Fischer, M.; Tenbohlen, S. The breakdown voltage of insulation oil under the influences of humidity, acidity, particles and pressure. In Proceedings of the International Conferences APTADM Wroclaw, Poland, 26–28 September 2007; pp. 1–7.

39. Wang, X.L.; Wang, Z.D. Particle effect on breakdown voltage of mineral and ester based transformer oils. In Proceedings of the Conference on Electrical Insulation and Dielectric Phenomena (CEIDP), Quebec, QC, Canada, 26–29 October 2008; pp. 598–602.

40. Kasahara, Y.; Kato, M.; Watanabe, S. Consideration on the relationship between dielectric breakdown voltage and water content in fatty acid esters. *J. Am. Oil Chem. Soc.* **2012**, *89*, 1223–1229. [CrossRef]

41. Zhou, Z.C. *High Voltage Technology*; Water Resources and Electric Power Press: Beijing, China, 1988.

42. Meng, Z.Y.; Yao, X. *Dielectric Physical Basis*; National Defense Industry Press: Beijing, China, 1980.

43. Yang, B.C.; Liu, X.B.; Dai, Y.S. *High Voltage Technology*; Chongqing University Press: Chongqing, China, 2002.

44. Hassanl, O.; Shyegani, A.A.; Borsi, H.; Gockenbach, E.; Abu-Elzahab, E.M.; Gilany, M.I. Detection of oil-pressboard insulation aging with dielectric spectroscopy in time and frequency domain measurements. In Proceedings of the International Conference on Solid Dielectrics, Toulouse, France, 5–9 July 2004; pp. 665–668.

45. Saha, T.K.; Purkait, P. Investigation of polarization and depolarization current measurements for the assessment of oil-paper insulation of aged transformers. *IEEE Trans. Dielectr. Electr. Insul.* **2004**, *11*, 144–154. [CrossRef]

46. Fofana, I.; Hemmatjou, H.; Farzaneh, M. Low temperature and moisture effects on polarization and depolarization currents of oil-paper insulation. *Electr. Power Syst. Res.* **2010**, *80*, 91–97. [CrossRef]

47. Paraskevas, C.D.; Vassiliou, P.; Dervos, C.T. Temperature dependent dielectric spectroscopy in frequency domain of high-voltage transformer oils compared to physicochemical results. *IEEE Trans. Dielectr. Electr. Insul.* **2006**, *13*, 539–546. [CrossRef]

48. Yuan, Q. Research on Experiment and Simulation of Frequency Domain Dielectric Response for Transformer Oil-Paper Insulation Aging. Master's Thesis, Chongqing University, Chongqing, China, 2010.

49. Hao, J. Study on Time/Frequency Domain Dielectric Spectroscopy and Space Charge Characteristics of Transformer Oil-Paper Insulation Thermal Aging. Ph.D. Thesis, Chongqing University, Chongqing, China, 2012.

50. Chen, J.D.; Liu, Z.Y. *Dielectric Physics*; China Machine Press: Beijing, China, 1988; pp. 146–160.

51. Zhou, Y.; Hao, M.; Chen, G.; Wilson, G.; Jarman, P. Study of the charge dynamics in mineral oil under a non-homogeneous field. *IEEE Trans. Dielectr. Electr. Insul.* **2015**, *22*, 2473–2482. [CrossRef]

52. Zaengl, W.S. Applications of dielectric spectroscopy in time and frequency domain for HV power equipment. *IEEE Electr. Insul. Mag.* **2003**, *19*, 9–22. [CrossRef]

53. Saha, T.K.; Purkait, P. Investigations of temperature effects on the dielectric response measurements of transformer oil-paper insulation system. *IEEE Trans. Power Deliv.* **2008**, *23*, 252–260. [CrossRef]

54. Ma, Z.Q. Frequency Domain Dielectric Response Characteristics and Condition Evaluation Method of Transformer Oil-Paper Insulation. Master's Thesis, Chongqing University, Chongqing, China, 2012.

55. Yang, F.B. Quantitative Evaluation Resarch on Moisture Content of Oil-Paper and Aging Degree of Transformer Insulation. Master's Thesis, Southwest Jiaotong University, Sichuan, China, 2017.

56. *Determination of Volume Resistivity of Power Oils*; Industrial Standard, DL/T421-1991; Standardization Administration of the People's Republic of China: Beijing, China, 1991.

57. Frood, D.G.; Gallagher, T.J. Space-charge dielectric properties of water and aqueous electrolytes. *J. Mol. Liq.* **1996**, *69*, 183–200. [CrossRef]

58. Zhou, Y.; Hao, M.; Chen, G.; Wilson, G.; Jarman, P. Study of the dielectric response in mineral oil using frequency-domain measurement. *J. Appl. Phys.* **2014**, *115*, 124105. [CrossRef]

59. Zhou, Y.; Hao, M.; Chen, G.; Wilson, G.; Jarman, P. Quantitative study of electric conduction in mineral oil by time domain and frequency domain measurement. *IEEE Trans. Dielectr. Electr. Insul.* **2015**, *22*, 2601–2610. [CrossRef]

60. Zhou, L.J.; Li, X.L.; Duan, Z.C.; Wang, X.J.; Gao, B.; Wu, G.N. Influence of cellulose aging on characteristics of moisture diffusion in oil-paper insulation. *Proc. CSEE* **2014**, *34*, 3541–3547.

61. Krins, M.; Borsi, H.; Gockenbach, E. Influence of carbon particles on the breakdown voltage of transformer oil. In Proceedings of the 12th International Conference on Conduction and Breakdown in Dielectric Liquids, Rome, Italy, 15–19 July 1996; pp. 296–299.

62. Hosier, I.L.; Ma, H.; Vaughan, A.S. Effect of electrical and thermal ageing on the breakdown strength of silicone oil. In Proceedings of the 18th IEEE International Conference on Dielectric Liquids (ICDL), Bled, Slovenia, 29 June–3 July 2014; pp. 160–163.

63. Hosier, I.L.; Vaughan, A.S. Effect of particulates on the dielectric properties and breakdown strength of insulation oil. In Proceedings of the 2017 Electrical Insulation Conference (EIC), Baltimore, MD, USA, 11–14 June 2017; pp. 376–379.

64. Lewis, T.D. Transactionson Dielectrics and Electrical Insulation. *IEEE Nanometr. Dielectr.* **1994**, *1*, 812. [CrossRef]

65. Choi, S.U.S. Enhancing thermal conductivity of fluids with nano-particles. *ASME Fed.* **1995**, *231*, 99–105.

66. Zmarz, Y.D.; Dobry, D. Analysis of properties of aged mineral oil doped with C60 fullerenes. *IEEE Trans. Dielectr. Electr. Insul.* **2014**, *21*, 1119–1126. [CrossRef]

67. Aksamit, P.; Zmarzly, D.; Boczar, T.; Szmechta, M. Aging properties of fullerene doped transformer oils. In Proceedings of the Conference Record of the 2010 IEEE International Symposium on Electrical Insulation, San Diego, CA, USA, 6–9 June 2010; pp. 1–4.

68. Prasath, R.T.A.R.; Roy, N.K.; Mahato, S.N.; Thomas, P. Mineral oil based high permittivity $CaCu_3Ti_4O_{12}$ (CCTO) nanofluids for power transformer application. *IEEE Trans. Dielectr. Electr. Insul.* **2017**, *24*, 2344–2353. [CrossRef]

69. Du, Y.F.; Lv, Y.Z.; Zhou, J.Q.; Li, X.X.; Li, C.R. Breakdown properties of transformer oil-based TiO_2 nanofluid. In Proceedings of the Conference of Electrical Insulation and Dielectric Phenomena, West Lafayette, IN, USA, 17–20 October 2010; pp. 1–4.

70. Du, Y.F.; Lv, Y.Z.; Wang, F.C. Effect of TiO$_2$ nanoparticles on the breakdown strength of transformer oil. In Proceedings of the Conference Record of the 2010 IEEE International Symposium on Electrical Insulation, San Diego, CA, USA, 6–9 June 2010; pp. 1–4.

71. Wang, Q.; Rafiq, M.; Lv, Y.; Li, C.; Yi, K. Preparation of Three Types of Transformer Oil-Based Nanofluids and Comparative Study on the Effect of Nanoparticle Concentrations on Insulating Property of Transformer Oil. *J. Nanotechnol.* **2016**, *2016*, 5802753. [CrossRef]

72. Du, Y.F.; Lv, Y.Z.; Zhou, J.Q.; Chen, M.T.; Li, X.X.; Li, C.R. Effect of ageing on insulating property of mineral oil-based TiO$_2$ nanofluids. In Proceedings of the International Conference on Dielectric Liquids, Trondheim, Norway, 26–30 June 2011; pp. 1–4.

73. Rafiq, M.; Li, C.; Lv, Y. Breakdown characteristics of transformer oil based silica nanofluids. In Proceedings of the Multi-Topic Conference, Islamabad, Pakistan, 5–6 December 2016; pp. 1–4.

74. Ma, J.; Zhou, Y.M.; Zhu, Z.G. Oxidation Resistance Study of Nano-Particles Modified Mineral Transformer Oil. *Electrotech. Electr.* **2016**, *10*, 47–51.

75. Katiyar, A.; Dhar, P.; Nandi, T. Effects of nanostructure permittivity and dimensions on the increased dielectric strength of nano insulating oils. *Colloids Surf. A Physicochem. Eng. Asp.* **2016**, *509*, 235–243. [CrossRef]

76. Yang, Q.; Liu, M.N.; Sima, W.X.; Jin, Y. Effect of electrode materials on the space charge distribution of an Al$_2$O$_3$ nanomodifed transformer oil under impulse voltage conditions. *J. Phys. D Appl. Phys.* **2017**, *50*, 46–48. [CrossRef]

77. Zhou, J.Q.; Du, Y.F.; Chen, M.T. AC and lightning breakdown strength of transformer oil modified by semiconducting nanoparticles. In Proceedings of the International Conference of Electrical Insulation and Dielectric Phenomena, Cancun, Mexico, 16–19 October 2011; pp. 652–654.

78. Peppas, G.D.; Charalampakos, V.P.; Pyrgioti, E.C. Statistical investigation of AC breakdown voltage of nanofluids compared with mineral and natural ester oil. *IET Sci. Meas. Technol.* **2016**, *10*, 644–652. [CrossRef]

79. Pugazhendhi, S.C. Experimental evaluation on dielectric and thermal characteristics of nano filler added transformer oil. In Proceedings of the International Conference on High Voltage Engineering and Application, Shanghai, China, 17–20 September 2012; pp. 207–210.

80. Liu, J.; Zhou, L.; Wu, G.; Zhao, Y.; Liu, P.; Peng, Q. Dielectric frequency response of oil-paper composite insulation modified by nanoparticles. *Proc. CSEE* **2011**, *19*, 510–520.

81. Cavallini, A.; Karthik, R.; Negri, F. The effect of magnetite, graphene oxide and silicone oxide nanoparticles on dielectric withstand characteristics of mineral oil. *IEEE Trans. Dielectr. Electr. Insul.* **2015**, *22*, 2592–2600. [CrossRef]

82. Srinivasan, C.; Saraswathi, R. Nano-oil with high thermal conductivity and excellent electrical insulation properties for transformers. *Curr. Sci.* **2012**, *102*, 1361–1363.

83. Shukla, G.; Aiyer, H. Thermal conductivity enhancement of transformer oil using functionalized nanodiamonds. *IEEE Trans. Dielectr. Electr. Insul.* **2015**, *22*, 2185–2190. [CrossRef]

84. Peña, C.D.L.; Rivera-Solorio, C.I.; Payán-Rodríguez, L.A. Experimental analysis of natural convection in vertical annuli filled with AlN and TiO$_2$/mineral oil-based nanofluids. *Int. J. Therm. Sci.* **2017**, *111*, 138–145. [CrossRef]

85. Lv, Y.; Du, Q.; Wang, L. Effect of TiO$_2$ nanoparticles on the ion mobilities in transformer oil-based nanofluid. *AIP Adv.* **2017**, *7*, 105022. [CrossRef]

86. You, Z. Influence Mechanism of Nanoparticles on the Insulating Properties of Nanofluid/Pressboard under Lighting Impulse Voltage. Ph.D. Thesis, North China Electric Power University, Hebei, China, 2015.

87. Hemmer, M.; Julliard, Y.; Badent, R.; Schwab, A.J. Streamer inception and propagation in rape-seed oils and mineral oils. In Proceedings of the Conference on 2001 Report Electrical Insulation and Dielectric Phenomena, Kitchener, ON, Canada, 14–17 October 2001; pp. 548–551.

88. Yusnida, M.; Azmi, K.; Ahmad, M.A.; Kamarol, M. Breakdown Strength Characteristic of RBDPO and Mineral Oil Mixture as an Alternative Insulating Liquid for Transformer. *J. Phys. Soc. Jpn.* **2013**, *64*, 1756–1759.

89. Rapp, K.J.; Mcshane, C.P.; Luksich, J. Interaction mechanisms of natural ester dielectric fluid and Kraft paper. In Proceedings of the IEEE International Conference on Dielectric Liquids, Coimbra, Portugal, 26 June–1 July 2005; pp. 393–396.

90. Liao, R.J.; Hao, J.; Liang, S.W.; Zhu, M.Z.; Yang, L.J. Influence of water and acid on the thermal aging of mineral oil mixed with natural ester oil-paper insulation. *Trans. China Electrotech. Soc.* **2010**, *25*, 31–37.

91. Hao, J.; Liao, R.J.; Yang, L.J.; Liang, S.W.; Yin, J.G. Spectra and thermogravimetric characteristics of the mineral oil mixed with natural ester oil-paper insulation thermal aging. *High Volt. Eng.* **2010**, *36*, 926–931.

92. Liao, R.J.; Yin, J.G.; Yang, L.J.; Liang, S.W.; Hao, J. XPS study on the inhibition mechanism of newly developed mixed insulating oil on the thermal aging of insulating paper. *Trans. China Electrotech. Soc.* **2011**, *26*, 136–142.

93. Liao, R.; Hao, J.; Yang, L.; Grzybowski, S. Study on aging characteristics of mineral oil/natural ester mixtures-paper insulation. In Proceedings of the IEEE International Conference on Dielectric Liquids, Trondheim, Norway, 1–4 June 2011.

94. Liao, R.J.; Zhang, S.; Yang, L.J.; Liu, B.; Hao, J. Comparative analyses of thermal aging and power frequency breakdown characteristics between natural ester paper insulation and mineral oil-paper insulation. *High Volt. Eng.* **2012**, *38*, 769–775.

95. Koopal, L.K.; Lee, E.M.; Bohmer, M.R. Adsorption of cationic and anionic surfactants on charged metal-oxide surfaces. *J. Colloid Interface Sci.* **1995**, *170*, 85–97. [CrossRef]

96. Allen, M.P.; Tildesley, D.J. *Computer Simulation of Liquids*; Clarendon Press: Oxford, UK, 1987; pp. 1–2.

97. Scamehorn, J.F.; Schechter, R.S.; Wade, W.H. Adsorption of surfactants on mineral oxide surfaces from aqueous-solutions.1. Isomerically pure anionic surfactants. *J. Colloid Interface Sci.* **1982**, *85*, 463–478. [CrossRef]

98. Fuerstenau, D.W. Equilibrium and nonequilibrium phenomena associated with the adsorption of ionic surfactants at solid-water interfaces. *J. Colloid Interface Sci.* **2002**, *256*, 79–90. [CrossRef]

99. Zhu, W.P. Application of molecular simulation technology to macromolecule. *Plast. Sci. Technol.* **2002**, *5*, 23–25.

100. Matthew, N.; Cristian, L.; Abhash, N.; Michael, T.K. Monte Carlo simulation of complex reaction systems: Molecular structure and reactivity in modelling heavy oils. *Chem. Eng. Sci.* **1990**, *45*, 2083–2088.

101. Stokke, B.T.; Christensen, B.E.; Smidsroed, O. Degradation of multistranded polymers: Effects of interstrand stabilization in xanthan and scleroglucan studied by a Monte Carlo method. *Macromolecules* **2002**, *25*, 2209–2214. [CrossRef]

102. Balazs, A.C.; Zhou, Z.; Yeung, C. Behavior of amphiphilic comb copolymers in oil/water mixtures: A molecular dynamics study. *Langmuir* **1992**, *8*, 2295–2300. [CrossRef]

103. Karaborni, S.; O'Connell, J.P. Molecular dynamics simulations of model micelles. 4. Effects of chain length and head group characteristics. *J. Phys. Chem.* **1990**, *94*, 2624–2631. [CrossRef]

104. Li, Q.M.; Huang, X.W.; Liu, T.; Yan, J.Y.; Wang, Z.D.; Zhang, Y.; Lu, X. Application progresses of molecular simulation methodology in the area of high voltage insulation. *Trans. China Electrotech. Soc.* **2016**, *31*, 1–13.

105. Lu, Y.C. Study on Diffusion Behavior of Gas and Mechanics of Oil-Paper Aging using Molecular Simulation. Ph.D. Thesis, Chongqing University, Chongqing, China, 2007.

106. Liang, S.W. Biodegradability of environment-friendly anti-aging transformer oil and its effect on thermal aging of insulating paper. *Electrotech. Appl.* **2011**, *30*, 74–78.

107. Liao, R.J.; Zhu, M.Z.; Yang, L.J.; Zhou, X.; Yan, J.M.; Sun, C.X. Analysis of interaction between transformer oil and cellulosic insulation paper using molecular simulation method. *High Volt. Eng.* **2011**, *37*, 268–275.

108. Liao, R.J.; Wang, K.; Yin, J.G.; Yang, L.J.; Sun, H.G.; D, X.P. Influence of initial moisture on thermal aging characteristics of oil-paper insulation. *High Volt. Eng.* **2012**, *38*, 1172–1178.

109. Zhang, Z.Q.; Yang, K.F.; Zhang, J.L. ReaxFF molecular dynamics simulations of non-catalytic pyrolysis of triglyceride at high temperatures. *RSC Adv.* **2013**, *3*, 6401–6407. [CrossRef]

110. Zhang, Y. Research on Micro-Mechanism of Pyrolysis Process of Oil-Paper Insulation in Power Transformer. Ph.D. Thesis, North China Electric Power University, Beijing, China, 2016.

111. Du, L.; Wang, W.J.; Chen, W.G. Reaction mechanism of transformer oil pyrolysis based on TG-DSC and molecular simulation. In Proceedings of the IEEE International Conference on High Voltage Engineering Application, Pilsen, Czech Republic, 6–8 September 2016; pp. 1–4.

112. Liu, J.; Wu, G.N.; Zhou, L.J.; Wen, Y.J.; Lv, H. Moisture diffusion in oil-paper insulation using molecular simulation. *High Volt. Eng.* **2010**, *36*, 2907–2912.

113. Yang, L.J.; Qi, C.L.; Wu, G.L.; Liao, R.J.; Wang, Q.; Gong, C.Y.; Gao, J. Molecular dynamics simulation of diffusion behaviour of gas molecules within oil-paper insulation system. *Mol. Simul.* **2013**, *39*, 988–999. [CrossRef]

114. Yeckel, C.A.; Curry, R.D. Electrostatic field simulation study of nanoparticles suspended in synthetic insulating oil. *IEEE Trans. Plasma Sci.* **2010**, *38*, 2514–2519.

115. Dai, J.Z.; Dong, M.; Wen, F.X.; Wang, L.; Ren, M. The molecular dynamic simulation investigation of the dispersion stability of nano-modified transformer oil. In Proceedings of the Conference on Electrical Insulation and Dielectric Phenomena, Ann Arbor, MI, USA, 18–21 October 2015; pp. 475–478.

116. Shi, J.; Luo, B.; Sima, W.X.; Cai, H.S. Experiment and simulation of the effect of nanoparticles on oil-paper composite insulation system. *South. Power Syst. Technol.* **2015**, *9*, 51–56.

117. Adil, L.; Jacqueline, L.S.; Yang, K.; Ren, G.Y. A molecular dynamic investigation of viscosity and diffusion coefficient of nanoclusters in hydrocarbon fluids. *Comput. Mater. Sci.* **2015**, *99*, 242–246.

118. Zhou, Z.J.; Jin, Y.; Kong, H.Y.; Fang, F.X. Simulation study on heat transfer characteristic of nano-modified transformer oil. *Insul. Mater.* **2016**, *49*, 19–23.

119. Yang, L.J.; Liao, R.J.; Sun, C.X.; Yin, J.G.; Zhu, M.Z. Influence of vegetable oil on the thermal aging rate of kraft paper and its mechanism. *Trans. China Electrotech. Soc.* **2012**, *38*, 381–384.

120. Li, Y.; Dong, M.; Dai, J.Z.; Zhou, J.R.; Wu, Z.Y.; Wang, J.Y. Molecular dynamics simulation on impact of temperature on viscosity of nano-modified transformer oil. *Insul. Mater.* **2017**, *10*, 1–11.

121. Lu, X.C.; Yu, J.; Alberto, S. Nanoparticles adsorbed at the water/oil interface: Coverage and composition effects on structure and diffusion. *Langmuir* **2013**, *29*, 7221–7228. [CrossRef] [PubMed]

122. Shi, P.; Zhang, H.; Lin, L.; Song, C.H. Molecular dynamics simulation of four typical surfactants at oil/water interface. *J. Dispers. Sci. Technol.* **2017**, *31*, 1–8. [CrossRef]

123. Passover, M.; Rostov, P. Molecular dynamic simulation of thermal conductivity of electrically insulating thermal nano-oil. *ASME Int. Mech. Eng. Congr. Expos.* **2012**, *40*, 1565–1571.

124. Ranatunga, R.; Nguyen, C.; Chiu, C.; Shinoda, W.; Nielsen, S. Molecular dynamics simulations of nanoparticles and surfactants at oil/water interfaces. *ACS Symp.* **2015**, *18*, 295–314.

125. Hosier, I.L.; Vaughan, A.S.; Swingler, S.G. Studies on the ageing behavior of various synthetic and natural insulation oils. In Proceedings of the IEEE International Conference on Dielectric Liquids, Futuroscope-Chasseneuil, France, 1–4 June 2008; pp. 1–4.

126. Hosier, I.L.; Vaughan, A.S.; Montjen, F.A. Ageing of biodegradable oils for high voltage insulation systems. In Proceedings of the IEEE Conference on Electrical Insulation and Dielectric Phenomena, Kansas City, MO, USA, 15–18 October 2006; pp. 481–484.

127. Li, X.H.; Li, J.; Sun, C.X.; Dang, J.L.; Li, Y. Study on electrical aging lifetime of vegetable oil-paper insulation. *Proc. CSEE* **2007**, *27*, 18–22.

128. Hosier, I.L.; Guushaa, A.; Westenbrink, E.W.; Rogers, C.; Vaughan, A.S.; Swingler, S.G. Aging of biodegradable oils and assessment of their suitability for high voltage applications. *IEEE Trans. Dielectr. Electr. Insul.* **2011**, *18*, 728–738. [CrossRef]

129. Hosier, I.L.; Rogers, C.; Vaughan, A.S.; Swingler, S.G. Ageing behavior of vegetable oil blends. In Proceedings of the 2010 Annual Report Conference on Electrical Insulation and Dielectric Phenomena, Southampton, UK, 1–4 October 2010; pp. 1–4.

130. Li, X.H.; Li, J.; Du, L.; Yang, L.J.; Sun, C.X. The electric properties of a transgenic vegetable oil. *Proc. CSEE* **2006**, *26*, 95–99.

131. Wilhelm, H.M.; Stocco, G.B.; Batista, S.G.J. Reclaiming of in-service natural ester-based insulating fluids. *IEEE Trans. Dielectr. Electr. Insul.* **2013**, *20*, 128–134. [CrossRef]

132. Singha, S.; Asano, R.; Frimpong, G.; Claiborne, C.C.; Cherry, D. Comparative aging characteristics between a high oleic natural ester dielectric liquid and mineral oil. *IEEE Trans. Dielectr. Electr. Insul.* **2014**, *21*, 149–158. [CrossRef]

energies

MDPI

Article

Experimental Study on Breakdown Characteristics of Transformer Oil Influenced by Bubbles

Chunxu Qin [1,*], Yan He [1], Bing Shi [1], Tao Zhao [2], Fangcheng Lv [2] and Xiangrui Cheng [2]

[1] School of Electrical Engineering, Northeast Electric Power University, No. 169 Changchun Road, Jilin 132012, China; yanhe1180@163.com (Y.H.); shibing2018@163.com (B.S.)

[2] Hebei Provincial Key Laboratory of Power Transmission Equipment Security Defense, North China Electric Power University, No. 619 North of Yonghua Street, Baoding 071003, China; alibabazhao@163.com (T.Z.); lfc@ncepu.edu.cn (F.L.); chengxiangrui94@163.com (X.C.)

* Correspondence: qincx@neepu.edu.cn

Received: 30 January 2018; Accepted: 9 March 2018; Published: 13 March 2018

Abstract: Bubbles will reduce the electric strength of transformer oil, and even result in the breakdown of the insulation. This paper has studied the breakdown voltages of transformer oil and oil-impregnated pressboard under alternating current (AC) and direct current (DC) voltages. In this paper, three types of electrodes were applied: cylinder-plan electrodes, sphere-plan electrodes, and cone-plan electrodes, and the breakdown voltages were measured in both no bubbles and bubbles. The sphere-sphere electrodes were used to study the breakdown voltage of the oil-impregnated pressboard. The results showed that under the influence of bubble, the breakdown voltage of the cylinder-plan electrode dropped the most, and the breakdown voltage of the cone-plan electrode dropped the least. The bubbles motion was the key factor of the breakdown. The discharge types of the oil-impregnated pressboard were different with bubbles, and under DC, the main discharge type was flashover along the oil-impregnated pressboard, while under AC, the main discharge type was breakdown through the oil-impregnated pressboard.

Keywords: breakdown voltage; transformer oil; pressboard; bubbles; AC; DC

---

1. Introduction

Mineral transformer oil is the main dielectric liquid for engineering application, and the oil and pressboard insulation system is widely used in high voltage electric power equipment manufacturing [1]. High-voltage DC projects are being developed in China now [2], and the AC and DC high voltages will be applied to the insulation of the converter transformer. The primary insulation of the converter transformer is also oil and pressboard insulation [3]. Bubbles will decrease the electrical strength of the oil-pressboard insulation, and will cause a partial discharge in the oil, finally resulting in the insulation breaking down due to the accumulation of the bubbles. Therefore, the bubbles are a key factor that affects the breakdown voltages of the oil-pressboard insulation [4].

Since the liquid discharge is so complex, there is no physical interpretation for all time and space stages of the development of the liquid discharge [5]. There is also no absolutely pure oil in engineering. Due to manufacturing, transportation, and other reasons, there will be some bubbles and particles in the oil. The generation progress of moisture in the mineral oil-impregnated pressboard has been studied previously. The oil-impregnated pressboard contains moisture, which will be released to the oil generally and, especially under high temperatures, the moisture is released quickly from the pressboard into the oil in the form of gas bubbles [6]. Researchers have studied the bubble-generating mechanism in oil-pressboard insulation of transformers. Heinrichs [7] discussed gas-evolving mechanisms in oil, pressboard, and combinations of oil and pressboard. Some models approximating the hot-spot configuration were designed to study the critical mechanisms under conditions that were approaching service overloads. A maximum transformer overload temperature was recommended from the results of this study. Kaufmann and McMillen also performed similar studies [8]. Oommen [9] has given an initial mechanism of bubble evolution and an equation to obtain bubble evolution temperatures. Koch and Tenbohlen [10] have studied the formation of bubbles in oil-paper insulating systems. The results showed that bubbles greatly reduce the dielectric withstand strength of the oil. Water accelerates the aging of the oil-paper insulation, decreases its dielectric strength, and generates bubbles at high temperatures. The bubbles will be harmful to transformer operation if the water content in the pressboard is above 2%. Przybylek et al. [11–13] have studied that it is easier to generate the bubbles in ageing paper than in new paper at a high temperature. Perkasa et al. [14] have studied the evolution of bubbles in vegetable oil-impregnated pressboard.

The behavior of bubbles plays an important role in the breakdown of the dielectrics. Sharbaugh et al. [15] suggested that breakdown occurs in the bubbles and the bubbles may be produced by the electronic progress. Ogata et al. [16] have found that the bubbles' size will decrease with the strength of the electric field, and the generation of an electrohydrodynamic (EHD) flow of the liquid will affect not only the movement of the bubble, but also the bubble evolution mechanism. Hara et al. [17] have studied the thermal bubble deformation and indicated that the breakdown voltages are close to the bubble behavior with pulse voltages. Seok et al. [16] have indicated that the bubbles' behavior was affected by a 60 Hz electric field and pressure, and the groove generated the bubbles' stream. In [18], Hara and Kubuki have suggested that a partial discharge might occur in the bubble before breakdown, if the bubble size is as large as 0.5 mm. In [19], Hara et al. have studied that the gradient force and Maxwell stress strongly affect the bubble dynamics and bubble shape in the gap, and the electric forces that are applied on the bubble lead to a lesser effect of the bubble on the breakdown voltage.

From the above discussion, the oil-paper insulation can generate the bubbles, which will decrease the electrical strength of the insulation. Most studies have been aimed at the bubble-generating mechanism and bubble behavior in the discharge of liquid nitrogen and helium, yet, there are few breakdown characteristics of transformer oil with the bubbles. Therefore, in this work, breakdown voltages of transformer oil and the oil-impregnated pressboard with the bubbles were studied under 50 Hz AC and positive DC power supplies.

2. Experimental Methods

2.1. Experimental System

The experimental system included the power supply (50 Hz AC and DC), oscilloscope, vacuum test chamber, and high-speed camera, as shown in Figure 1. Where 1 is the 50 Hz power frequency testing transformer, 2 and 5 are protected resistors, 3 is a HV diode, 4 is a HV capacitor, 6 is the potential divider, 7 is the oscilloscope, 8 is a syringe, 9 is a bushing, 10 is the electrode system that was immersed in the oil tank with dimensions of 35 cm × 20 cm × 20 cm, 11 is the vacuum test chamber (the dimensions of the vacuum test chamber are 60 cm × 50 cm × 40 cm), and 12 is high-speed camera, and the model of the high-speed camera is FASTEC-IL5, produced by Rocky Mountain High Speed, with a frame rate of 1000 frames per second.

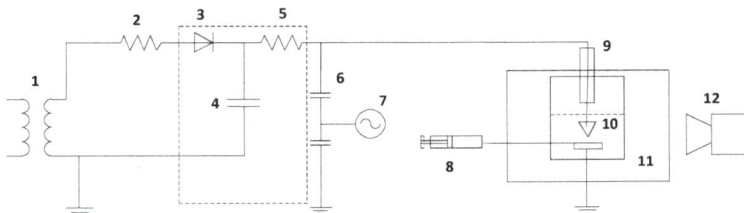

Figure 1. The schematic diagram of the experimental system.

The test oil was KI 25X transformer oil, and the main characteristics are shown in Table 1. The size of the oil-impregnated pressboard is 50 mm × 35 mm, and the thickness of the pressboard is 1 mm. The pressboard was soaked in the oil for more than 24 h, and then the oil in the pressboard reached the saturation state. From the Oommen curve [20], the water content of the oil-impregnated pressboard is 6%. The size of the electrodes is shown as Figure 2, and the material of the electrodes is brass. The distribution of the electric field will influence the bubbles' movement, so we designed the cone-plan electrodes to study the extremely non-uniform field, and as is known, in the transformer insulation, a more uniform electric field is the typical one, so we select the other two electrode systems (cylinder-plan electrodes and sphere-plan electrodes) in order to study the relative uniform electric fields. The sphere-sphere electrodes (a typical relative uniform electric field) was used for the investigation of the breakdown voltage of the pressboard.

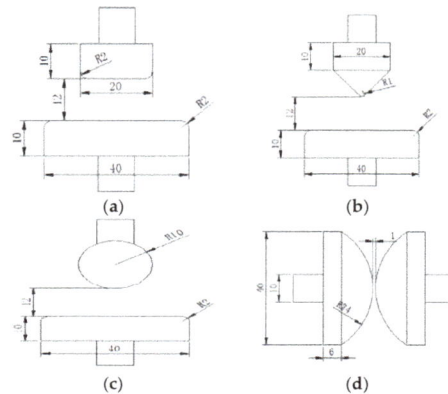

Figure 2. The schematic diagram of the electrodes (mm). (**a**) cylinder-plan electrodes; (**b**) cone-plan electrodes; (**c**) sphere-plan electrodes; and, (**d**) sphere-sphere electrodes used for the investigation of the breakdown voltage of the pressboard.

Table 1. Characteristics of the tested oils.

Dielectric Liquid	Moisture (ppm)	Viscosity (mm²/s), 40 °C	Flash Point (°C)	Tan δ (90 °C, 50 Hz)	Acid Number (mg·KOH/g)
KI25X	≤40	10.13	142	0.0006	0.05

2.2. Test Method

2.2.1. The Breakdown of the Oil

The tests were carried out at a high voltage experimental hall which is about 40 m × 20 m × 6 m, and the test method is as follows:

1. Before voltage was applied, about 7 L of initial oil was put into the electrode systems, and the oil's surface is 5 cm above the electrodes. After 30 min' standing, the bubbles were injected by a syringe at a speed of three bubbles per second, and the bubbles linked the two electrodes. The bubbles were injected all the time until the breakdown occurred.
2. Then, AC or DC voltages were applied on the electrodes at the rate of voltage rise of 1 kV/s until breakdown of the oil. We then waited 0.5 h.
3. We changed the oil and repeated the tests.
4. All of the discharges were recorded by the high-speed camera.

2.2.2. The Breakdown of the Oil-Impregnated Pressboard

The sphere-sphere electrodes were selected to study the breakdown voltages of the oil-impregnated pressboard. Before the tests, the oil-impregnated pressboard was put between the two electrodes and we injected the oil into the electrode system. The surface of the oil is one centimeter taller than the electrodes, so the entire discharge system was immersed in oil.

1. AC or DC voltages were applied on the electrodes and the rate of voltage rise was 1 kV/s until breakdown. We changed the pressboard, mixed the oil, and then waited 0.5 h.
2. Using a vacuum to reduce air pressure to 100 Pa, the bubbles appeared on the oil-impregnated pressboard. AC or DC voltages were applied to the electrodes and the rate of voltage rise was 1 kV/s until breakdown of the insulation.
3. We mixed the oil, changed the oil-impregnated pressboard, and repeated steps 1 and 2.
4. All of the discharges were recorded by the high-speed camera.

3. Results and Discussion

3.1. AC Breakdown Voltages of the Transformer Oil

From Figure 3a, the breakdown voltages of the cone-plan electrodes without bubbles are more than the breakdown voltages of the other two types electrodes, and the breakdown voltages of the sphere-plan electrodes are similar to the cylinder-plan electrodes. Dissado [21] and Peppas [22] indicated that the statistical analyses were suitable for the dielectric breakdown. Based on these, we calculated some statistics of the breakdown voltages.

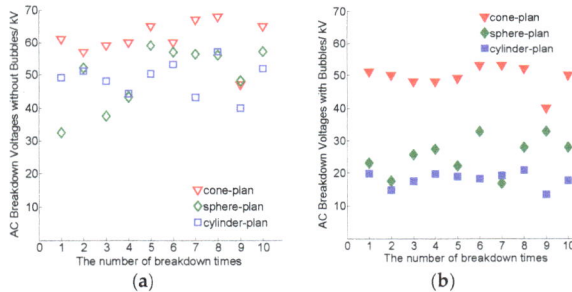

Figure 3. The AC breakdown voltages (effective values) of the transformer oil. (**a**) The AC breakdown voltages of the oil without bubbles; and, (**b**) the AC breakdown voltages of the oil with bubbles.

From Table 2, the average value of the breakdown voltages of the cone-plan electrodes is 60.9 kV, the average value of the breakdown voltages of the sphere-plan electrodes is 49.9 kV, and the average value of the breakdown voltages of the cylinder-plan electrodes is 48.8 kV.

It can be seen that the system with the most non-uniform field distribution had the greatest withstand voltage. As it was known, the breakdown theory of the transformer oil is a discharge bridge mechanism, and according to this theory, the breakdown voltage is determined by the formation of the bridge. Under the most uniform field distribution, there is a corona that will disturb the discharge bridge path, and it is more difficult to form the breakdown path than under the uniform fields. Therefore, the cone-plan electrodes have the highest breakdown voltages.

From Figure 3b, the breakdown voltages of the cone-plan electrodes with bubbles are much more than the breakdown voltages of the other two types of electrodes, and the breakdown voltages of the sphere-plan electrodes are more than the cylinder-plan electrodes. The average value of the breakdown voltages of the cone-plan electrodes is 49.4 kV, the average value of the breakdown voltages of the sphere-plan electrodes is 25.3 kV and the average value of the breakdown voltages of the cylinder-plan electrodes is 18.0 kV.

Table 2. Statistics of the AC breakdown voltages (BDV).

Electrode Systems		Cone-Plan	Sphere-Plan	Cylinder-Plan
With bubbles	Mean BDV (kV)	49.4	25.3	18
	Std. Deviation (kV)	3.8	5.5	2.3
	Min BDV (kV)	40	16.8	13.4
	Max BDV0 (kV)	53	32.8	20.8
Without bubbles	Mean BDV (kV)	60.9	49.9	48.8
	Std. Deviation (kV)	6.1	9.2	5.1
	Min BDV (kV)	47	32.5	40
	Max BDV (kV)	68	59	57.1

From the above, with the influence of the bubbles, the breakdown voltages of the cone-plan electrodes have a 18.9% decrease on average; the bubbles have led to a 49.1% decrease in the value of the average breakdown voltage for the sphere-plan electrodes; and, the bubbles have also led to a 62.3% decrease in the value of average breakdown voltage for the cylinder-plan electrodes. The results showed that the bubbles would decrease the breakdown voltages for all of the electrodes, and the reduction of the cylinder-plan electrodes was the most serious. Meanwhile, the cone-plan electrodes had the minimal effects.

The electric strength and electric field distribution affect the breakdown voltages and the progresses of the discharge. The three electrodes have different electric distributions: the electric fields of the cylinder-plan electrodes and the sphere-plan electrodes are more uniform than that of the cone-plan electrodes. When the bubbles were generated from the central of the plan electrode, and they formed a bridge that linked the two electrodes, and then the AC high voltage was applied on the electrodes. If there was a uniform electric field between the two electrodes, there was no corona that can disturb the bubbles bridge, and the discharge would take place in the bubbles bridge, and that would cause the breakdown of the two electrodes. However, if there was a non-uniform electric field between the two electrodes, then the corona would take place near the highest electric field area, which would disturb the bubbles and the length of the bridge would get extended, then the discharge path in the non-uniform electric field would be longer than that of the uniform electric field, the progress of which is shown as Figure 4. The breakdown voltages of the non-uniform electric field saw a more minor effect than that of the uniform electric field. It can be concluded that the electric field can affect the movements of the bubbles and change the discharge path, which directly affected the breakdown voltages of the two electrodes. Thus, the breakdown voltages of the cone-plan electrodes had a minimal effect, and the breakdown voltages of the other two types of electrodes had more serious effects by the bubbles. Since the uniformity of the cylinder-plan electrodes' electric field was close to the sphere-plan electrodes, the reductions of the breakdown voltages of the two types electrodes are close.

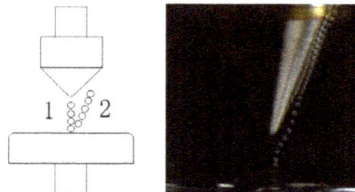

Figure 4. The bubbles of the cone-plan electrodes, where 1 are the initial bubbles; 2 are the bubbles affected by the electric field. The left picture is the schematic diagram and the right one is the real photo of the bubbles affected by the electric field.

3.2. Positive DC Breakdown Voltages of the Transformer Oil

From Figure 5a, the breakdown voltages of the three types of electrodes without bubbles were different from that of the AC conditions. From Table 3, the average value of the breakdown voltages of the cone-plan electrodes is 45.7 kV, the average value of the breakdown voltages of the sphere-plan electrodes is 56.7 kV, and the average value of the breakdown voltages of the cylinder-plan electrodes is 49.5 kV.

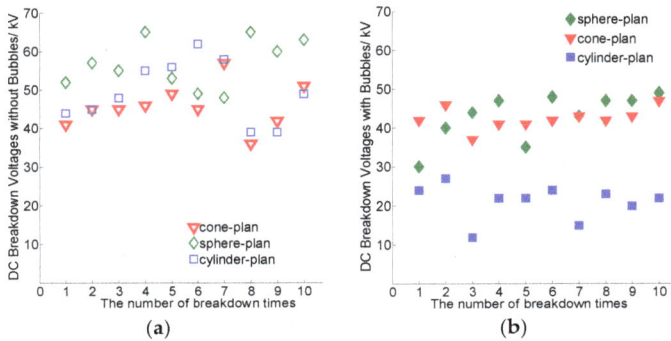

Figure 5. The positive DC breakdown voltages of the transformer oil. (**a**) The DC breakdown voltages of the oil without bubbles; (**b**) the DC breakdown voltages of the oil with bubbles.

From Figure 5b, the breakdown voltages of the cylinder-plan electrodes with bubbles are much lower than the breakdown voltages of the other two types of electrodes, and the breakdown voltages of the sphere-plan electrodes are close to that of the cone-plan electrodes. From Table 3, the average value of the breakdown voltages of the cone-plan electrodes is 42.4 kV, the average value of the breakdown voltages of the sphere-plan electrodes is 43.0 kV, and the average value of the breakdown voltages of the cylinder-plan electrodes is 21.0 kV.

From the above, with the influence of the bubbles, the breakdown voltages of the cone-plan electrodes have a 7.2% decrease on average; the bubbles have led to a 24.1% decrease in the value of The average breakdown voltage for the sphere-plan electrodes; and also the bubbles have led to a 66.9% decrease in the value of The average breakdown voltage for the cylinder-plan electrodes. The results showed that under a DC voltage, the bubbles would also decrease the breakdown voltages for all the electrodes, and the reduction of the cylinder-plan electrodes' breakdown voltages were the most serious, and the cone-plan electrodes had minimal effects. However, the sphere-plan electrodes saw a minor reduction than that of the AC condition.

From the above, the breakdown voltages' decrease of the cylinder-plan electrodes under positive DC voltage were similar to that under AC voltage, however, the decreases of the breakdown voltages of the sphere-plan and cone-plan electrodes are much smaller than those of the AC conditions.

From Section 3.1, it is can be seen that the discharge progress was influenced by the movements of the bubbles: if the bubbles were easy to link the two electrodes, the breakdown voltage would see a sharp decrease. Under the DC electric field, the bubbles were easy to charge, and they would then obtain an electric force with constant direction that made the bubbles move faster than under the AC condition. For that, the bubbles would move from the plan electrode to the other electrode with a higher speed under the DC condition, and this progress made it difficult for the bubbles to link the two electrodes, especially in the non-uniform electric fields, like the electric field of the cone-plan electrodes, the existence of the corona would make the bubbles' bridge more difficult to form.

From Figure 4, the bubbles in path 2 with high speed were easy to spread to other regions in the oil, which would make the bubbles difficult to link the electrodes, so the breakdown voltages had a minimal reduction, and in [15], the authors made a similar conclusion, that the strong electric field prevented the bubbles from flowing steadily. From Figure 6a, the central of the two electrodes formed a uniform electric field, and there was no corona and disturbance, the bubbles would move from the center of the plan electrode to the center of the cylinder electrode directly, like path 1, and there was little possibility to form path 2 under the uniform electric field. From Figure 6b, the bubbles of the sphere-plan electrodes also spread easily to other areas in the oil from the edge of the sphere, so it was also difficult to form the bubbles to link the two electrodes, and the decrease of breakdown voltages was minor compared to the AC condition. However, in the case of the cylinder-plan electrodes, the decrease of the breakdown voltages under the DC condition was approximately equal to that under the AC condition.

Table 3. Statistics of the AC breakdown voltages (BDV).

Electrode Systems		Cone-Plan	Sphere-Plan	Cylinder-Plan
With bubbles	Mean BDV (kV)	42.4	43	21.1
	Std. Deviation (kV)	2.7	6.2	4.4
	Min BDV (kV)	37	30	12
	Max BDV (kV)	47	49	27
Without bubbles	Mean BDV (kV)	45.7	56.7	49.5
	Std. Deviation (kV)	5.7	6.3	8.0
	Min BDV (kV)	36	48	39
	Max BDV (kV)	42.4	43	21.1

From the above analyses, it was difficult for the bubbles to link the two electrodes in the non-uniform electric field under the DC condition, so the breakdown voltages of the cone-plan electrodes were minimally affected by the bubbles. However, under the force of the DC electric field, there was little effect on the formation of the bubbles in the uniform electric field, so the descent of the breakdown voltage of the cylinder-plan electrodes under the DC condition was approximately equal to that under the AC condition.

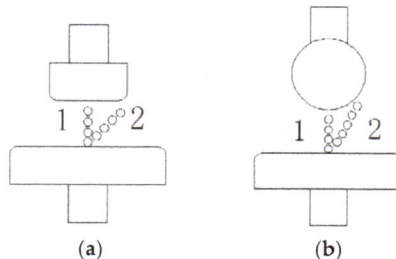

Figure 6. The bubbles of the cylinder-plan and sphere-plan electrodes, where 1 is the initial bubbles; 2 shows the bubbles affected by the corona; (**a**) the bubbles of the cylinder-plan electrodes; and, (**b**) the bubbles of the sphere-plan electrodes.

3.3. The Breakdown Voltages of the Oil-Impregnated Pressboard

From Figure 7a, under positive DC voltage, the breakdown voltages of the oil-impregnated pressboard have decreased by the bubbles produced by the oil-impregnated pressboard. From Table 4, the average value of the breakdown voltages of the oil-impregnated pressboard without bubbles was 29.5 kV, and the average breakdown voltage with bubbles was 15.1 kV. This means that with the influence of the bubbles, the breakdown voltages of oil-impregnated pressboard have a 48.9% decrease, on average.

From Figure 7b, under AC voltage, the breakdown voltages of the oil-impregnated pressboard have decreased by the bubbles that are produced by the oil-impregnated pressboard. In Table 4, the average breakdown voltages of the oil-impregnated pressboard without bubbles was 22.1 kV, and the average breakdown voltage with bubbles was 16.8 kV. This means that with the influence of the bubbles, the breakdown voltages of oil-impregnated pressboard have a 25% decrease, on average.

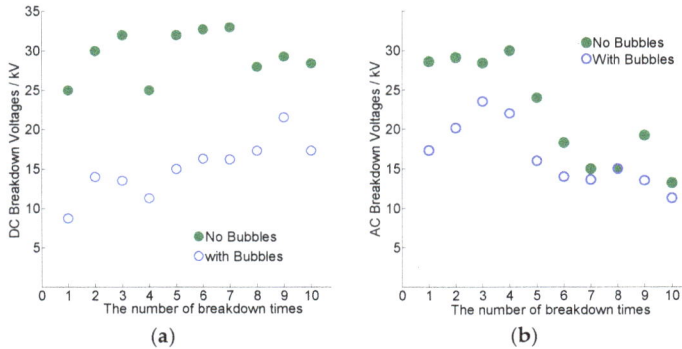

Figure 7. The breakdown voltages of the oil-impregnated pressboard. (**a**) The DC breakdown voltages; and (**b**) the AC breakdown voltages (effective values).

Table 4. Statistics of the oil-impregnated pressboard breakdown voltages (BDV).

Electrode Systems		AC	AC*	DC
With bubbles	Mean BDV (kV)	16.6	19	15.1
	Std. Deviation (kV)	4.0	3.1	3.5
	Min BDV (kV)	11.3	16	8.7
	Max BDV (kV)	23.5	23.5	21.5
Without bubbles	Mean BDV (kV)	22.8	28	29.5
	Std. Deviation (kV)	6.6	2.3	2.9
	Min BDV (kV)	13.2	29.1	25
	Max BDV (kV)	30	16	33

However, it can be seen that there was a sharp decline of the breakdown voltages after about ten discharges, in total. The breakdown voltages of the oil-impregnated pressboard with bubbles have no obvious reduction when compared with the breakdown voltages without bubbles. Thus, the data before the sharp decline were selected again to calculate the average breakdown voltages, shown as AC* (in Table 4), the average value without bubbles was 28 kV, and the value with bubbles was 19.8 kV. There was a 29.4% decrease of the average breakdown voltage under the influence of the bubbles that were produced by the oil-impregnated pressboard.

The AC breakdown voltages showed a sharp decline after about nine or ten breakdown discharges. The breakdown progress damaged the oil-impregnated pressboard and the oil, and some fiber, carbon, and other products can decrease the electric strength of the oil. Under the AC condition, the breakdown progress had multiple discharges (shown as Figure 8), and the oil and the pressboard decomposed into other chemical compositions that would decrease the electric strength of the oil, and that the compositions had a cumulative effect. It can be concluded that the cumulative effect in our test will play an important role after about 10 AC breakdown discharges. On the other hand, under the DC condition, the DC discharge only had one discharge (seen in Figure 8), and this caused slight damage to the pressboard and the oil, and the repeated breakdown tests can change the morphology of water in oil, which make the breakdown voltage of transformer oils increase. In [23], they reached a similar conclusion. Thus, the DC breakdown voltages are larger for the last tests than the first.

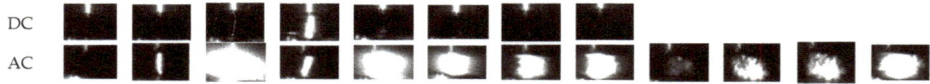

Figure 8. The breakdown progress of the cylinder-plan electrodes in the oil.

Table 5 shows the discharge forms of the oil-impregnated pressboard under the DC and AC electric fields. The breakdown meant that when the discharge occurred, the oil impregnated pressboard was broken down by the high voltage, and, meanwhile, the flashover meant that when the discharge occurred, the oil-impregnated pressboard was not broken down, but the discharge occurred along the oil-impregnated pressboard, only breaking down the bubbles and oil. From Table 5, under the AC voltage, bubbles or no bubbles, the discharge forms were mainly breakdown, and under the DC condition, the discharge forms with no bubbles were breakdown. However, the main discharge forms with bubbles under the DC condition have changed into flashover.

Table 5. The discharge forms of the oil-impregnated pressboard under AC and DC electric fields.

Voltages	AC		DC	
Test No.	No Bubbles	Bubbles	No Bubbles	Bubbles
1	Breakdown	Breakdown	Breakdown	Flashover
2	Breakdown	Breakdown	Breakdown	Flashover
3	Breakdown	Breakdown	Breakdown	Flashover
4	Breakdown	Breakdown	Breakdown	Flashover
5	Breakdown	Breakdown	Breakdown	Flashover
6	Breakdown	Flashover	Breakdown	Flashover
7	Breakdown	Breakdown	Breakdown	Flashover
8	Breakdown	Breakdown	Breakdown	Flashover
9	Breakdown	Breakdown	Breakdown	Breakdown
10	Breakdown	Breakdown	Breakdown	Flashover

From the above results, with the influence of bubbles, the descent of the breakdown voltages under DC voltage was much more than that under AC voltage; this is because the main discharge forms under DC voltage with bubbles were flashover, and that had a minor breakdown voltage.

In DC, the discharge almost takes place in the bubbles and the insulation cannot be recovered, thus the bubble will move up to the top surface of the oil along the paper, so the discharge occurred in the bubbles along the paper surface, and the flashover was formed. However, in AC, on one hand, the insulation of the bubbles can be recovered after breakdown, so the discharge will be shut down when the bubbles move toward up to the top, while, on the other hand, the oil-impregnated paper has a greater dielectric loss and produce more heat, which will reduce the insulation strength of the paper. Additionally, in AC, the pressboard is a complex insulation, there are oil and air in the paper, and there will be a higher electric strength on them, thus making them easier to breakdown, but this is not a problem in DC, and the authors in [24] share the same opinion. Therefore, the discharge form of the pressboard is always that of breakdown in AC.

4. Conclusions

The presence of the bubble will reduce the breakdown voltages of the oil-paper insulation. From the above discussion, the work may be concluded as follows:

In AC and DC, the bubbles decreased the breakdown voltage of the cylinder-plan electrodes the most among the three electrodes systems, and the mean values have dropped more than 60%, whereas the cone-plan electrodes have the least decline. That is say the bubbles have the minimum effect on the non-uniform electric field. This is due to the movement of the bubbles and the corona will prevent the bubbles' discharge path. We also found that for the corona preventing the fibers and particles to form

Energies **2018**, *11*, 634

the bridge, in AC, the cone-plan electrodes have the largest breakdown voltages in the oil without bubbles. In DC, the decrease of the breakdown voltages of the cone-plan and sphere-plan electrodes influenced by the bubbles were much less than that in AC.

For the oil-impregnated pressboard, the decrease of the breakdown voltage under DC voltage was more than that under AC voltage. In our work, the main discharge form is flashover in DC, however, the main discharge form is breakdown through the pressboard. The phenomenon is determined by the discharge mechanisms in DC and AC. The discharge is the single discharge in DC, which cannot destroy the oil-paper insulation, but it can change the morphology of water in oil, and which can cause the BDV increase. Meanwhile, the AC breakdown is a kind of multiple high intensity discharge, which can seriously destroy the pressboard and oil, and the cumulative effect of fibrinogen degradation products will reduce the BDV.

Author Contributions: Chunxu Qin, Tao Zhao and Fangcheng Lv conceived and designed the experiments; Yan He, Bing Shi and Xiangrui Cheng performed the experiments; Chunxu Qin and Tao Zhao analyzed the data; Fangcheng Lv contributed reagents/materials/analysis tools; Chunxu Qin wrote the paper.

Conflicts of Interest: The authors declare no conflict of interest.

References

1. Liao, R.; Ying, L.; Zheng, H.; Ma, Z. Reviews on oil-paper insulation thermal aging in power transformers. *Trans. China Elctrotech. Soc.* **2012**, *27*, 1–12.
2. Yin-biao, S. Present status and prospect of HVDC transmission in China. *High Volt. Eng.* **2004**, *30*, 1–2.
3. You-liang, S.; Qing-pu, W.; Wen-ping, L. R&D ofconverter transformer for ±800 kV HVDC power transmission project. *Electr. Equip.* **2006**, *7*, 17–20.
4. Tao, Z.; Yunpeng, L.; Fangcheng, L.; Ruihong, D. Study on dynamics of the bubble in transformer oil under non-uniform electric field. *IET Sci. Meas. Technol.* **2016**, *10*, 498–504. [CrossRef]
5. Ushakov, V.Y.; Klimkin, V.F.; Korobeynikov, S.M. *Impulse Breakdown of Liquids*, 1st ed.; Springer: Berlin/Heidelberg, German; GmbH & Co. K: Berlin, German, 2007; ISBN 9783642091858.
6. Heinrichs, F.W. Bubble formation in power transformer windings at overload temperatures. *IEEE Trans. Power Appar. Syst.* **1979**, *PAS-98*, 1576–1582. [CrossRef]
7. Kaufmann, G.H.; McMillen, C.J. Gas bubble studies and impulse tests on distribution transformers during loading-above nameplate rating. *IEEE Trans. Power Appar. Syst.* **1983**, *PAS-102*, 2531–2542. [CrossRef]
8. Oommen, T.V.; Lindgren, S.R. Bubble evolution from transformer overload. *Transm. Distrib. Conf. Expo.* **2001**, *1*, 137–142.
9. Koch, M.; Tenbohlen, S. Evolution of bubbles in oil-pressboard insulation influenced by material quality and ageing. *IET Electr. Power Appl.* **2011**, *5*, 168–174. [CrossRef]
10. Przybylek, P. The influence of cellulose insulation aging degree on its water sorption properties and bubble evolution. *IEEE Trans. Dielectr. Electr. Insul.* **2010**, *17*, 906–912. [CrossRef]
11. Przybylek, P.; Nadolny, Z.; Moscicka-Grzesiak, H. Bubble effect as a consequence of dielectric losses in cellulose insulation. *IEEE Trans. Dielectr. Electr. Insul.* **2010**, *17*, 913–919. [CrossRef]
12. Przybylek, P.; Moranda, H.; Walczak, K.; Moscicka-Grzesiak, H. Can the bubble effect occur in an oil-impregnated pressboard bushing. *IEEE Trans. Dielectr. Electr. Insul.* **2012**, *19*, 1879–1883. [CrossRef]
13. Perkasa, C.Y.; Nick, L.; Tadeusz, C.; Jaury, W.; Daniel, M. A Comparison of the Formation of Bubbles and Water Droplets in Vegetable and Mineral Oil Impregnated Transformer pressboard. *IEEE Trans. Dielectr. Electr. Insul.* **2014**, *21*, 2111–2118. [CrossRef]
14. Sharbaugh, H.; Devins, J.S.; Rzad, S.J. Progress in the field of electric breakdown in dielectric liquids. *IEEE Trans. Electr. Insul.* **1978**, *13*, 249–276. [CrossRef]
15. Ogata, S.; Shigehara, K.; Yoshida, T.; Shinohara, H. Small bubble formation by using strong non-uniform electric field. *IEEE Trans. Ind. Appl.* **1980**, *IA-16*, 766–770. [CrossRef]
16. Hara, M.; Koishihara, H.; Saita, K. Breakdown behavior of cryogenic liquids in the presence of thermal bubbles under ramped voltage. *IEEE Trans. Dielectr. Electr. Insul.* **1991**, *26*, 685–691. [CrossRef]
17. Seok, B.Y.; Tamuro, N.; Hara, M. A study of thermal bubble behavior in the simulated electrode system of HT superconducting coils. *IEEE Trans. Dielectr. Electr. Insul.* **1999**, *6*, 109–116. [CrossRef]

18. Hara, M.; Kubuki, M. Effect of thermally induced bubbles on the electrical breakdown characteristics of liquid nitrogen. *Proc. IEE* **1990**, *137*, 209–216.

19. Hara, M.; Wang, Z.; Saito, H. Thermal Bubble Breakdown in Liquid Nitrogen under Nonuniform Fields. *IEEE Trans Dielectr. Electr. Insul.* **1994**, *4*, 709–716. [CrossRef]

20. Gan, D.; Liu, F.; Liu, P.; Hu, C. Calculation Method of Oil-Paper Insulation Moisture Content Based on Oil-Paper Moisture Equilibrum of Transformer. *Transformer* **2009**, *8*, 21–24.

21. Dissado, L.A. Theoretical basis for the statistics of dielectric breakdown. *J. Phys. D Appl. Phys.* **1990**, *23*, 1582–1591. [CrossRef]

22. Peppas, G.D.; Charalampakos, V.P.; Pyrgioti, E.C.; Danikas, M.G.; Bakandritsos, A.; Gonos, I.F. Statistical investigation of AC breakdown voltage of nanofluids compared with mineral and natural ester oil. *IET Sci. Meas. Technol.* **2016**, *10*, 644–652. [CrossRef]

23. You, Z.; Jun, J.; Yingting, L.; Hao, W.; Yuzhen, L.; Chengrong, L. Statistical Analysis of AC Breakdown Voltage for Transformer Oil. *Insul. Mater.* **2015**, *48*, 73–77.

24. Wang, Y.; Wei, X.; Chen, Q. Breakdown Characteristics of Oil-pressboard Insulation in Converter Transformer under Composite Electric Field and Its Test Method. *High Volt. Eng.* **2011**, *37*, 2005–2011.

energies

MDPI

Article

Comparison of Positive Streamers in Liquid Dielectrics with and without Nanoparticles Simulated with Finite-Element Software

Juan Velasco [1,†]**, Ricardo Frascella** [1,†]**, Ricardo Albarracín** [1,*]**, Juan Carlos Burgos** [2]**, Ming Dong** [3]**, Ming Ren** [3] **and Li Yang** [3]

[1]   Departamento de Ingeniería Eléctrica, Electrónica, Automática y Física Aplicada, Escuela Técnica Superior de Ingeniería y Diseño Industrial (ETSIDI), Universidad Politécnica de Madrid (UPM), Ronda de Valencia 3, 28012 Madrid, Spain; juan.velasco.ares@alumnos.upm.es (J.V.); r.frascella@alumnos.upm.es (R.F.)

[2]   Electrical Engineering Department, Universidad Carlos III de Madrid, Avda. de la Universidad 30, 28911 Leganés, Madrid, Spain; jcburgos@ing.uc3m.es

[3]   State Key Laboratory of Electrical Insulation for Power Equipment, Xi'an Jiaotong University, Xi'an 710049, China; dongming@mail.xjtu.edu.cn (M.D.); renming@mail.xjtu.edu.cn (M.R.); aixiyuxin2012@stu.xjtu.edu.cn (L.Y.)

*   Correspondence: ricardo.albarracin@upm.es

†   These authors contributed equally to this work.

Received: 15 December 2017; Accepted: 29 January 2018; Published: 3 February 2018

Abstract: In this paper, a comparison of positive streamer diffusion propagation is carried out in three configurations of oil transformers: mineral transformer oil, mineral oil with solid dielectric barriers, and a nanofluid. The results have been solved using a finite-element method with a two-dimensional (2D) axi-symmetric space dimension selected. Additionally, previous results from other research has been reviewed to compare the results obtained. As expected, it is confirmed that the nanoparticles improve the dielectric properties of the mineral oil. In addition, it is observed that the dielectric solid blocks the propagation of the streamer when it is submerged with a horizontal orientation, thus perpendicular to the applied electric field. The computer used, with four cores (each 3.4 GHz) and 16 GB of RAM, was not sufficient for performing the simulations of the models with great precision. However, with these first models, the tendency of the dielectric behavior of the oil was obtained for the three cases in which the streamer was acting through the transformer oil. The simulation of these models, in the future, in a supercomputer with a high performance in terms of RAM memory may allow us to predict, as an example, the best concentration of nanoparticles to retard the streamer inception. Finally, other dielectric issues will be predicted using these models, such as to analyze the advantages and drawbacks of the presence of dielectrics inside the oil transformer.

Keywords: streamer; transformer oil; nanoparticles; liquid insulation; solid insulation; Comsol

1. Introduction

Transformers are an important part of the electric power system, because they allow us to modify the voltage of electric energy. Transformers make it possible to combine a high transport voltage (which helps to reduce transport losses) with a low voltage in the consumption (necessary to guarantee the safety of the users).

Electrical insulation is the main component of electrical devices such as power transformers, power cables, engines and electric generators. In transformers, different types of insulation systems, such as solids and liquids, which are widely used in transformers, can be found.

The most widely used dielectric liquid in power transformers is mineral oil. This is because it has better insulating properties and a higher thermal conductivity than gaseous insulation systems

[1]. In addition, it is auto-regenerated in the case of short-discharges, which means that it is not significantly deteriorated by partial discharge (PD) activity, unlike solid insulation.

Transformer oil is occasionally submitted to abnormally high levels of electric field. If these levels are maintained for a long enough time, it can produce ionization of the oil molecules, which can lead to a streamer, which is a precursor of the electric arc that provokes the total dielectric breakdown of a dielectric. Generally, the mean velocity of positive streamer propagation is about 10 times larger than that for the negative needle in oil liquid insulation for power transformers [2]. For this reason, positive applied voltages can lead to positive streamers with a greater risk for the insulating fluids than negative streamers, as a result of the dielectric breakdown that occurs at lower voltage values and with shorter breakdown times than negative streamers [3,4].

During the previous years, the properties of insulating oils have been investigated to improve their dielectric and thermal behavior [1,5]. Lately, most research has been focused on identifying the advantages and disadvantages of adding nanoparticles into transformer oil [6]. The oil and nanoparticle assembly is commonly referred to as a nanofluid.

Nanotechnology is a field of sciences that is dedicated to the study, control and manipulation of elements whose dimensions are smaller than micrometers. Therefore, it is applied to scales of atomic and molecular level. Nanotechnology is widely used in the food, electronics or medicine industries.

Other key research aimed at improving the dielectric properties of oil focuses on the dielectric solids submerged in the oil. Experimental tests have demonstrated that submerged dielectric solids can facilitate or block the streamer diffusion. The present document studies the influence of dielectric solids perpendicular to the direction of the streamer. This is because of the fact that this configuration blocks the diffusion of the streamer, whereas the position parallel to the streamer facilitates its progress [7].

In addition to the experimental measurements, the use of finite-element software allows for a faster advance in the study of the results obtained for the dielectric rupture of transformer oil [7,8].

O'Sullivan [8] presented his PhD thesis on the development of streamers in mineral transformer oil. Then, Hwang [7] improved the O'Sullivan model by increasing the breakdown voltage, adding nanoparticles and immersed solid barriers. Later, Jadidian [9] improved the propagation model of the streamer, studied the effect of different immersed dielectrics and constructed three-dimensional (3D) models for simulating streamer branches.

This paper focuses on the comparison of positive streamers in three different systems: mineral oil, nanofluids and immersed dielectric solids in mineral oil. For the study of the cases mentioned above, simulations have been implemented in Comsol Multiphysics using the electrodynamic model developed by researchers at the Massachusetts Institute of Technology (MIT) and ABB Corporate Research [7–9]. The geometry used for the simulations corresponded to the International Electrotechnical Commission (IEC)-60897 [10], with a gap of 25 mm. The voltage was applied in the form of a step with a rise time of 10 ns.

The paper is structured as follows: Section 2 explains the geometric characteristics of the tip-sphere electrodes corresponding to the IEC-60897 [10]. Section 3 exposes the theoretical basis of the models implemented to simulate the cases mentioned above. The results for the simulations obtained are shown in Section 4, the challenges for the streamer simulation are described in Section 5, and the conclusions are presented in Section 6.

2. Oil Test Specimen

The geometry used in the realization of the simulations is included in the IEC-60897, *Methods for the determination of the lightning impulse breakdown voltage of insulating liquids* [10].

The configuration consists of a cell with two electrodes. The first has a spherical tip with a radius of between 40 and 70 µm, and the second has a spherical shape with a diameter of between 12.5 and 13 mm. The spherical electrode is connected to the ground and a potential is applied to the spherical tip electrode. The separation between the electrodes can vary between 10 and 25 mm, and the cell must contain around 300 mL of oil. The geometry of the simulated cell is shown in Figure 1.

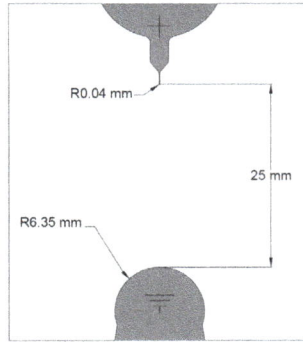

Figure 1. Geometry corresponds to International Electrotechnical Commission (IEC)-60897 [10].

3. Streamer Model

Several studies have shown that the key phenomena in the development of streamers is the direct ionization of the oil molecules, as a result of the action of the electric field. When high levels of electric field are reached, the ionization of the oil takes place, creating free charge carriers. Another source of charge are the generated electrons coming from the cathode as a result of the electric field. This creates a flow of electrons from the cathode to the anode. This flow of charges creates a conductive path called a streamer [11].

There are different mechanisms that take place in the ionization process of transformer oil. These methods have been studied by O'Sullivan [8] and Hwang [7], who have concluded that the most important mechanism that participates in the ionization of the oil is the molecular ionization, dependent on the electric field, the "field ionization".

The models developed by O'Sullivan [8], Hwang [7], and Jadidian [9] are used as a basis for the development of the three case studies. For modeling the streamer, Jadidian [12] used the computer configuration shown in the first column of Table 1 at the MIT, while the configuration that has been used in this study at Escuela Técnica Superior de Ingeniería y Diseño Industrial (ETSIDI) is that shown in the second column. As depicted in Table 1, the MIT's computer has approximately 12 times more RAM than the computer used at ETSIDI, which allowed us to implement 7.2×10^6 degrees of freedom instead of 2×10^6 degrees of freedom, allowing for a more accurate model and requiring a time simulation close to 20 h, four times longer than the model implemented with the ETSIDI's configuration. The following sections explain the models that allowed us to perform the simulations and the subsequent comparison of the obtained results.

Table 1. Computers used for streamer simulation in oil transformers at Massachusetts Institute of Technology (MIT) and at Escuela Técnica Superior de Ingeniería y Diseño Industrial (ETSIDI).

Item	MIT	ETSIDI HP ProDesk 400 G3
Number of computers (used in parallel)	3	1
Number of cores	36 (3.4 GHz each)	4 (3.4 GHz each)
RAM	188 GB	16 GB
Degrees of freedom	$\approx 7.2 \times 10^6$	$\approx 2 \times 10^5$
Time of simulation	≈ 20 h	≈ 5 h

3.1. Streamers in Mineral Transformer Oil

The equations that control the behavior and interaction between the oil particles used for the simulations of the oil ionization and afterward the propagation of a streamer were developed by O'Sullivan [8] and Hwang [7].

The most widely used equation among the scientific community for modeling the field ionization ratio is based on the Zener tunneling model [13], and the mathematical expression of the density of the ionization field is as follows:

$$G_I(|\vec{E}|) = \frac{e^2 n_0 \alpha |\vec{E}|}{h} exp\left(-\frac{\pi^2 m^* \alpha \Delta^2}{eh^2 |\vec{E}|}\right) \tag{1}$$

where $\vec{E}$ (V/m) is the module of the electric field vector, h (m^{2}kgs^{-1}) is Planck's constant, e (C) is the charge of the electron, m^* (kg) is the mass of the electron in the ionization proces, α (m) is the intermolecular distance, n_0 (m^{-3}) is the density of ionizable species, and Δ (eV) is the electrical potential needed for the molecular ionization. Jadidian [9] found that the last four above-mentioned parameters are the most difficult to determine, because they depend on the chemical nature of the oil.

It could be said that almost all of the parameters are constant. Only the electric field is variable, because it depends on the geometry and the applied voltage.

To model the propagation and behavior of the free charge carriers in mineral oil, continuity equations have been used. The free charges in the oil are positive ions (Equation (2)), negative ions (Equation (3)), and electrons (Equation (4)). At the metal-liquid interface, positive ions and electrons are generated, but negative ions are not. Therefore, the electrons are injected from the electrode into the liquid, generating in turn positive ions, while the generation mechanism of negative ions is different because they are generated later by collisions, as depicted in [7,8].

$$\frac{\partial \rho_p}{\partial t} + \nabla \cdot (\rho_p \mu_p \vec{E}) - \frac{\rho_p \rho_e R_{pe}}{e} - \frac{\rho_p \rho_n R_{pn}}{e} = G_I(|\vec{E}|) \tag{2}$$

$$\frac{\partial \rho_e}{\partial t} - \nabla \cdot (\rho_e \mu_e \vec{E}) + \frac{\rho_p \rho_e R_{pe}}{e} - \frac{\rho_e}{\tau_a} = -G_I(|\vec{E}|) \tag{3}$$

$$\frac{\partial \rho_n}{\partial t} - \nabla \cdot (\rho_n \mu_n \vec{E}) + \frac{\rho_p \rho_n R_{pn}}{e} - \frac{\rho_e}{\tau_a} = 0 \tag{4}$$

where ρ_p (c/m^3), ρ_n (c/m^3) and ρ_e (c/m^3) are the space charge density of positive and negative ions and electrons, respectively; μ_e (m^{2}V^{-1}s^{-1}), μ_n (m^{2}V^{-1}s^{-1}) and μ_p (m^{2}V^{-1}s^{-1}) are the mobility of electrons and negative and positive ions, respectively. R_{pn} (m^{3}s^{-1}) and R_{pe} (m^{3}s^{-1}) are the ion-ion and electron-ion recombination ratios for the oil, respectively.

The three continuity equations are linked by the Poisson equation [14]. With this equation, it is possible to relate the variation of the electrical potential to the density of free charges in the oil. The Poisson equation is the following:

$$\nabla \cdot (\epsilon_1 \vec{E}) = \rho_p + \rho_n + \rho_e \tag{5}$$

where $\vec{E} = -\nabla V$ and ε_1 is the relative permittivity of the oil.

During the propagation of the streamer, there is heat dissipation due to the Joule effect. The oil temperature is related to the Joule heating through the product $E \cdot J$, where J (Am^{-2}) is the total current density. These equations are as follows:

$$\frac{\partial T}{\partial t} + \vec{v} \cdot \nabla T = \frac{1}{\rho_l c_v}(k_T \cdot \nabla^2 T + \vec{E} \cdot \vec{J}) \tag{6}$$

$$\vec{J} = (\rho_p \mu_p - \rho_e \mu_e - \rho_n \mu_n)\vec{E} \tag{7}$$

where ρ_l is the oil density, k_T (W/mK^{-1}) is the thermal conductivity of the oil, c_v (Jkg^{-1}K^{-1}) is the specific heat, T (K) is the temperature of the oil and $\vec{v} = 0$ (m/s) is the velocity of the oil, which can be considered negligible because of the fact that the streamer propagation time is on the scale of microseconds to nanoseconds.

The values of previously mentioned parameters are set out in Table 2 [8].

Table 2. Parameters used in the simulations for oil characterization.

Symbol	Parameter	Value	Unit
Δ	Ionization electrical potential	7.1	eV
n_0	Density of ionizable species	1×10^{23}	m^{-3}
α	Intermolecular distance	3×10^{-10}	m
e	Charge of the electron	-1.602×10^{-19}	C
m^*	Effective mass of the electron	9.1×10^{-32}	kg
R_{pe}, R_{pn}	Ion-ion and electron-ion recombination ratios	1.645×10^{-17}	m^3s^{-1}
μ_n, μ_p	Mobility of negative and positive ions	10^{-9}	$m^2V^{-1}s^{-1}$
μ_e	Mobility of electrons	10^{-4}	$m^2V^{-1}s^{-1}$
c_v	Specific heat	1.7×10^3	$Jkg^{-1}K^{-1}$
ρ_l	Oil density	880	kgm^{-3}
k_T	Thermal conductivity	0.13	W/mK^{-1}
τ_a	Electron attachment time constant	200×10^{-9}	s
ε	Vacuum permittivity	8.854×10^{-12}	F/m
h	Planck's constant	6.63×10^{-34}	m^2kgs^{-1}

3.2. Streamer in Nanofluid

The latest technological advances in this field have focused on improving the resistance to rupturing and the thermal behavior of the insulation. In this section, the effect of adding nanoparticles to the dielectric oils on the streamer formation is analyzed. These dielectrics have rupture voltages higher than for mineral oils; this is because the electrons are attracted to the nanoparticles, remaining attached to them and preventing their free movement [15].

In this model, it has been assumed that the nanoparticles have a perfect conductivity; when a potential affects them, they are perfectly polarized. For this to be possible, the electrons have to be deposited uniformly on the surface of the nanoparticle.

The effect of attracting electrons to the nanoparticle is called charge. This process takes place until the surface is completely covered with electrons; when this happens the nanoparticle is saturated and cannot attract more electrons. Therefore, charge only occurs in areas where the radial component of the electric field at the surface is positive. The equation governing the charge of the nanoparticle is the following [7]:

$$cos(\Theta) = -\frac{Q(t)}{12\pi\varepsilon_1 E^0 R^2} \tag{8}$$

where ε_1 (F/m) is the oil's permittivity, E^0 (V/m) is the electric field, R is the radius of the nanoparticle and $Q(t)$ is the time-dependent function of the nanoparticle charge. The saturation charge is given when $cos(\Theta)$ has a value of 1; therefore the formula of saturation charge is the following [7]:

$$Q_S = -12\pi\varepsilon_1 E^0 R^2 \tag{9}$$

Another important parameter to characterize the nanoparticle is the time constant. The time constant is the time during which the nanoparticle influences the electrodynamic development of the streamer. The equation for obtaining the time constant of the nanoparticles is the following [7]:

$$\tau_{np} = \frac{4\varepsilon_1}{|\rho_e|\mu_e} \tag{10}$$

The mobility of the electrons is higher than that of the ions; therefore, the nanoparticle captures them faster. By this fact, the nanoparticle becomes a negative charge in motion. The equation that determines the movement of the nanoparticle in the oil is the following [7]:

$$\mu_{np} = \frac{|Q_S|}{6\pi\eta R} \tag{11}$$

where the kinetic viscosity is $\eta = 0.002$ Pa $\cdot$ s.

The development of the model with nanofluids is based on that described for the propagation of streamers in mineral oil. These equations have to be slightly modified to include the influence of the nanoparticles. The controlling equations of the model are the following [16]:

$$G_I(|\vec{E}|) = \frac{e^2 n_0 \alpha |\vec{E}|}{h} exp\left(-\frac{\pi^2 m^* \alpha \Delta^2}{eh^2 |\vec{E}|}\right) \tag{12}$$

$$\frac{\partial \rho_p}{\partial t} + \nabla \cdot (\rho_p \mu_e \vec{E}) - \frac{\rho_p \rho_e R_{pe}}{e} - \frac{\rho_p \rho_n R_{pn}}{e} - \frac{\rho_p (\rho_n + \rho_{np}) R_{pn}}{e} = G_I(|\vec{E}|) \tag{13}$$

$$\frac{\partial \rho_e}{\partial t} - \nabla \cdot (\rho_e \mu_e \vec{E}) + \frac{\rho_p \rho_e R_{pe}}{e} - \frac{\rho_e}{\tau_a} + \frac{\rho_e}{\tau_{np}} \left(1 - H\left(\rho_{np,sat} - \rho_{np}\right)\right) = -G_I(|\vec{E}|) \tag{14}$$

$$\frac{\partial \rho_n}{\partial t} - \nabla \cdot (\rho_n \mu_n \vec{E}) + \frac{\rho_p \rho_n R_{pn}}{e} - \frac{\rho_e}{\tau_a} = 0 \tag{15}$$

$$\frac{\partial \rho_{np}}{\partial t} - \nabla \cdot (\rho_{np} \mu_{np} \vec{E}) + \frac{\rho_p \rho_{np} R_{pn}}{e} - \frac{\rho_e}{\tau_{np}} \left(1 - H\left(\rho_{np,sat} - \rho_{np}\right)\right) - \frac{\rho_p \rho_{np} R_{pn}}{e} = 0 \tag{16}$$

$$\nabla \cdot (\epsilon_1 \vec{E}) = \rho_p + \rho_n + \rho_e + \rho_{np} \tag{17}$$

$$\frac{\partial T}{\partial t} + \vec{v} \cdot \nabla T = \frac{1}{\rho_l c_v} (k_T \cdot \nabla^2 T + \vec{E} \cdot \vec{J}) \tag{18}$$

where ρ_p (cm^{-3}), ρ_n (cm^{-3}) and ρ_e (cm^{-3}) are the espace charge density of positive and negative ions and electrons, respectively; μ_e (m^{2}V^{-1}s^{-1}), μ_n (m^{2}V^{-1}s^{-1}) and μ_p (m^{2}V^{-1}s^{-1}) are the mobility of electrons and negative and positive ions, respectively. R_{pn} (m^{3}s^{-1}) and R_{pe} (m^{3}s^{-1}) are the ion-ion and electron-ion recombination ratios for oil, respectively; τ_a (s) is the electron attachment time constant, ρ_l (kgm^{-3}) is the oil density, K_t (W/mK^{-1}) is the thermal conductivity of the oil, C_v (Jkg^{-1}K^{-1}) is the specific heat, T is the temperature of the oil, $\vec{v} = 0$ (m/s) is the velocity of the oil, τ_{np} (s) is the time constant of nanoparticles and $H(\rho_{np,sat} - \rho_{np})$ is the step function that determines the saturation of the nanoparticles. The current density is also affected by the nanoparticles; the corresponding equation is $\vec{J} = (\rho_p \mu_p - \rho_e \mu_e - \rho_n \mu_n - -\rho_{np} \mu_{np})\vec{E}$ [7].

The saturation of the nanoparticles is determined by the $H(\rho_{np,sat} - \rho_{np})$ function; thus it is necessary to determine the number of electrons that are able to be attracted to its surface. The first step is to determine the volume of nanoparticles in the fluid. To do this, it is necessary to compare the magnetic saturation of the fluid $\mu_0 M_S(T)$ with the saturation of the nanoparticles $\mu_0 M_d(T)$ [7]:

$$\phi = \frac{M_S}{M_d} \tag{19}$$

The maximum charge density of electrons that can be deposited on the surface of the nanoparticle, whose volume is V_{np} (m^3), is the following [7]:

$$\rho_{np,sat}^{\infty} = 11 electrons \cdot \frac{-e}{V_{np}/\phi} \tag{20}$$

The values of the new parameters used in the model with nanofluids are set out in the Table 3 [7]:

Table 3. Values of the new parameters used in the streamer simulation for nanofluids.

Symbol	Parameter	Value	Unit
μ_{np}	Mobility of charged nanoparticles	$\sim 1 \times 10^{-9}$	$m^2V^{-1}s^{-1}$
$\rho_{np,sat}$	Nanoparticle charge density upper limit	500	cm^{-3}
Q_S	Saturation charge	-1.836×10^{-18}	C
R	Radius of nanoparticles	5	nm
μ_0	Vacuum permeability	$4\pi \times 10^{-7}$	H/m
$\mu_0 M_S$	Saturation magnetization of nanofluid	10^{-4}	T
$\mu_0 M_d$	Domain magnetization of nanoparticles	0.56	T
V_{np}	Nanoparticle volume	$\approx 5.236 \times 10^{-25}$	m^3
π	Volume fraction of nanoparticles	10^{-4}	-
$\rho_{np,sat}^{\infty}$	Upper limit for catching electrons	-600	cm^{-3}

The nanofluid has been modeled using nanoparticles with a radius of 5 nm and a maximum load of $\rho_{np,sat}^{\infty} = 500$ cm^{-3}; this is because the charge of the nanoparticles reaches values close to 80%, and therefore a value at infinity is not adequate [7]. A magnetization domain of 0.56 T has been considered. It has also been assumed that the nanoparticles have perfect conductivity, that is, when an electric field falls on them, they are perfectly polarized.

3.3. Streamer in Oil with Dielectric Barrier

One way of improving the insulation properties of a liquid is to immerse a dielectric solid oriented perpendicular to the direction of propagation of the streamer. During its propagation, the streamer stumbles upon the solid dielectric, impeding its advance [7]. This dielectric solid enables the liquid to withstand an applied voltage higher than that obtained for mineral oil.

Because of the difference in permittivity between the solid dielectric and the oil, there is an attraction of free charge carriers toward the solid because of the force of polarization. This creates a path of spatial density of free charge carriers accumulating on the surface of the solid dielectric where the streamer would propagate.

In this paper, simulations have been performed for three different cases of immersed dielectric solids. These three solid dielectrics have different relative permittivities; therefore, it has been possible to study the effect of permittivity differences on streamer propagation. The solid dielectrics that have been used are the following: polytetrafluoroethylene (PTFE), polyethylene terephthalate (PET) and Pressboard, whose relative permittivities are 2.2, 3.3 and 4.4, respectively.

To model the liquid-immersed dielectric solid system, some equations have been added to those indicated above in Section 3.1. These are Equations (21) and (22), which govern the submerged dielectric solids, and Equation (23), which models the dielectric liquid–solid contact zone. These equations were developed by Hwang [7] and Jadidian [9].

$$\nabla \cdot (\epsilon_r \epsilon_0 \vec{E}) = 0 \tag{21}$$

$$\vec{J^s} = 0 \tag{22}$$

$$\frac{\partial \rho_s}{\partial t} = \vec{n} \cdot (\rho_p \mu_p - \rho_e \mu_e - \rho_n \mu_n) \vec{E} \tag{23}$$

where ϵ_r is the relative permittivity of the solid dielectric; ϵ_0 is the vacuum permittivity, whose value is 8.8542×10^{-12} (C^2/Nm^2); J_s is the current density; and ρ_s is the surface charge density that appears on the contacting surface between the dielectrics.

The geometry used in this case is that shown in Figure 2. It consists of the geometry depicted in the IEC-60897 [10] standard, maintaining a gap of 25 mm between the electrodes, but with the dielectric solid placed 0.5 mm from the tip of the positive electrode (z-axis).

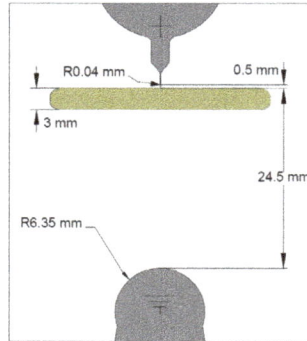

Figure 2. Solid dielectric immersed geometry.

4. Results of Positive Streamer Simulation

In this section, three models have been implemented for obtaining streamers in different oil transformer configurations. First, we describe the streamer in mineral transformer oil, in accordance with the model described in [8]. Second, we give the streamer in mineral transformer oil with nanoparticles, where we have implemented the model implemented in [7] with two different time constants of the nanoparticles, τ_{np} = 50 ns and τ_{np} = 20 ns, added. Finally, the streamer in mineral transformer oil with immersed dielectric solids has been simulated, in accordance with [9]. To allow for the comparison, the applied voltage was 300 kV for all the cases analyzed.

In the next two subsections, the following graphical representations are shown:

- Streamer propagation
- Net space charge
- Electric field
- Electric potential
- Temperature

In the third subsection, the graphical representation is presented of the streamer propagation at different time steps when three different solid dielectrics with permittivity values ϵ_r = 2.2, ϵ_r = 3.3 and ϵ_r = 4.4, respectively, were used.

4.1. Streamer in Mineral Transformer Oil

As can be seen in Figure 3, ions and electrons are attracted by the electric field between electrodes in such a way that the maxima of electrical charge are displaced. The net space charge density is shown in Figure 4a. In turn, this provokes a displacement of the electrical field, as shown in Figure 4b. The electric potential also changes with time (Figure 4c).

The region of maximum temperature is given by the points of space in which the ions are. As time passes, more free ions are produced, but while there is no avalanche phenomenon, the net number of charges formed is not high, and the temperature does not increase appreciably (Figure 4d). The increase in temperature is because the temperature is directly proportional to the electric field and the current density.

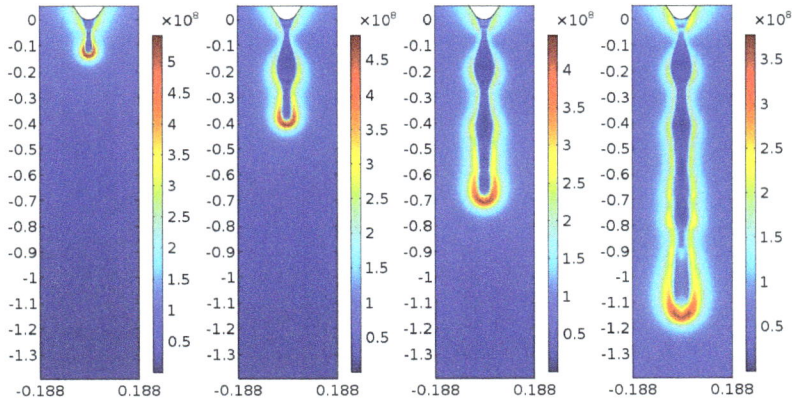

Figure 3. Positive streamer propagation. Evolution of the electric field (V/m) in mineral transformer oil along the z-axis, applying 300 kV and with a rise time of 10 ns.

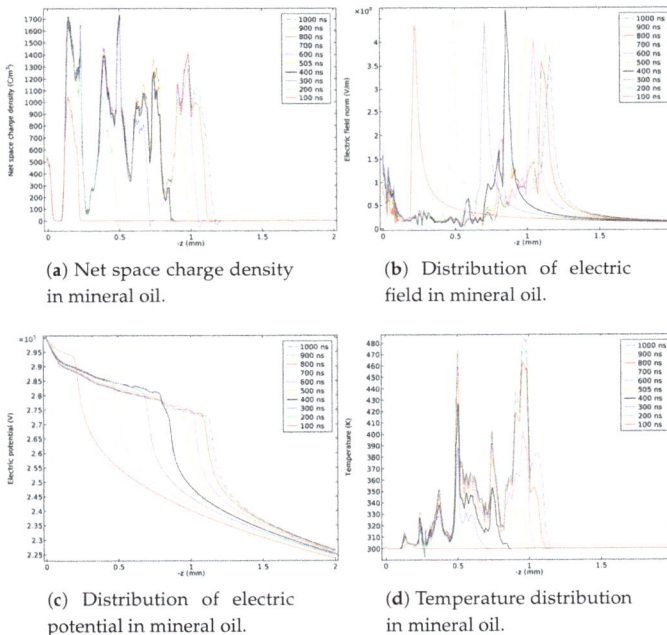

(a) Net space charge density in mineral oil.

(b) Distribution of electric field in mineral oil.

(c) Distribution of electric potential in mineral oil.

(d) Temperature distribution in mineral oil.

Figure 4. (a) Net space charge density, (b) distribution of electric field, (c) distribution of electric potential, and (d) temperature distribution along the z-axis from $t = 0$ to 1000 ns in intervals of 100 ns from the simulations in mineral transformer oil.

In mineral oil, a displacement of the electric field (Figure 4b) and the potential (Figure 4c) occurred as a result of the ionization of the oil. In accordance with real data presented in [17], an electric field level greater than 2×10^8 (V/m) is required to initiate a positive streamer in mineral oil. Therefore, the simulation results presented in this work are in agreement with those results presented through simulation in [8] and by real data in [17].

4.2. Streamer in Nanofluid

In order to study the influence of the time constant on the streamer process, in this subsection, simulations performed for two different time constants of τ_{np} = 50 ns and τ_{np} = 20 ns are discussed. These time constants of the nanoparticles added to the base fluids, indicating the time that it took for the nanoparticles to be charged.

4.2.1. Nanofluids with τ_{np} = 50 ns

When nanoparticles were added to the oil, the concentration of electrons was reduced. Therefore, the speed of the streamer propagation along the z-axis was reduced (Figure 5) compared with that in mineral oil (Figure 3). The space charge density was also somewhat reduced (Figure 6a) in the nanofluid compared with that in the base fluid, along the z-axis (Figure 4a). For this reason, the electric field advanced slowly (Figure 6b), and as did the electric potential (Figure 6c). Finally, it is shown in Figure 6d that the streamer temperature was also reduced.

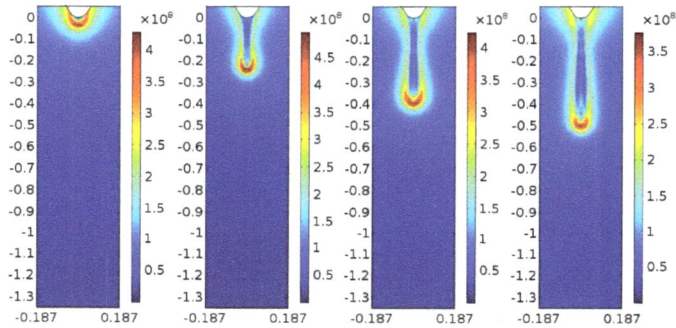

Figure 5. Positive streamer propagation. Evolution of the electric field (V/m) in nanofluids with τ_{np} = 50 ns along the z-axis, applying 300 kV and with a rise time of 10 ns.

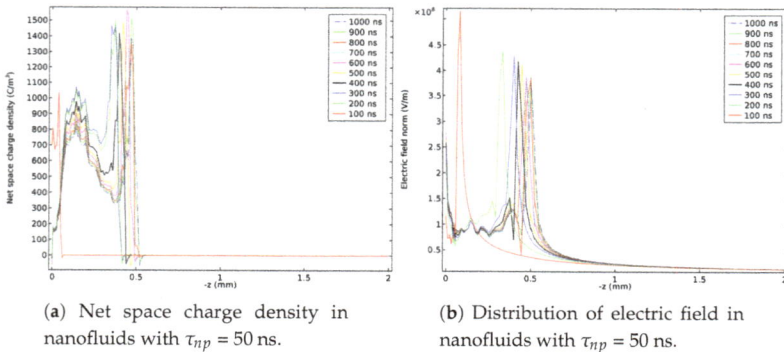

(**a**) Net space charge density in nanofluids with τ_{np} = 50 ns.

(**b**) Distribution of electric field in nanofluids with τ_{np} = 50 ns.

Figure 6. *Cont.*

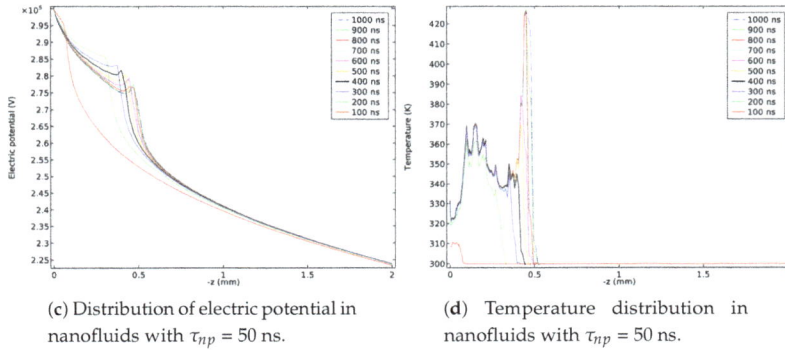

(**c**) Distribution of electric potential in nanofluids with τ_{np} = 50 ns.

(**d**) Temperature distribution in nanofluids with τ_{np} = 50 ns.

Figure 6. (**a**) Net space charge density, (**b**) distribution of electric field, (**c**) distribution of electric potential, and (**d**) temperature distribution along the z-axis from *t* = 0 to 1000 ns in intervals of 100 ns from the simulations in nanofluids with τ_{np} = 50 ns.

4.2.2. Nanofluids with τ_{np} = 20 ns

As the nanoparticle constant became smaller, the length of the streamer slightly decreased (Figure 7), and this was also observed for the net space charge density (Figure 8a). This was because the nanoparticles charged faster, which reduced the length of the streamer. The electric potential and temperature were also slightly reduced compared with the case of τ_{np} = 50 ns nanoparticles. Thus, when the time constant of the nanoparticles was reduced from τ_{np} = 50 ns to τ_{np} = 20 ns, this reduced the streamer speed and temperature; thus the probability of an electrical breakdown is diminished for an applied voltage of 300 kV, under these conditions.

In accordance with [7], it was shown that the addition of nanoparticles with a time constant of 50 ns could reduce the electric field from 6×10^2 V/m in mineral oil to 5.5×10^2 V/m in the nanofluid. Additionally, the streamer propagation velocity for the nanofluids was slower than that of the mineral transformer oil. These results are in agreement with those obtained experimentally in [18], where it is shown that the presence of conductive nanoparticles in mineral oil reduces the positive streamer propagation velocity. In [7], the model resulted in an average positive streamer velocity decrease from 1.65 km/s in mineral transformer oil to 1.05 km/s in the nanofluid; these are velocities very close to those obtained experimentally by [18]. In Table 4, a comparison is shown between simulation results obtained for positive streamer propagation in mineral oil and nanofluids when a voltage set at U = 300 kV is applied to the geometry shown in Figure 1. The positive streamer length obtained was 1.2 mm for mineral oil, instead of 1.6 mm presented in [8]. Additionally, the velocity for mineral oil was 1.2 km/s; this was 0.707 km/s for nanofluids with nanoparticles of τ_{np} = 50 ns and was 0.593 km/s for nanofluids with nanoparticles of τ_{np} = 20 ns. These differences were mainly because of the challenges for streamer simulation described in Section 5.

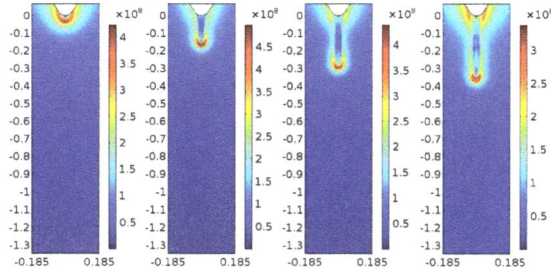

Figure 7. Positive streamer propagation. Evolution of the electric field (V/m) in nanofluids with τ_{np} = 20 ns along the z-axis, applying 300 kV and with a rise time of 10 ns.

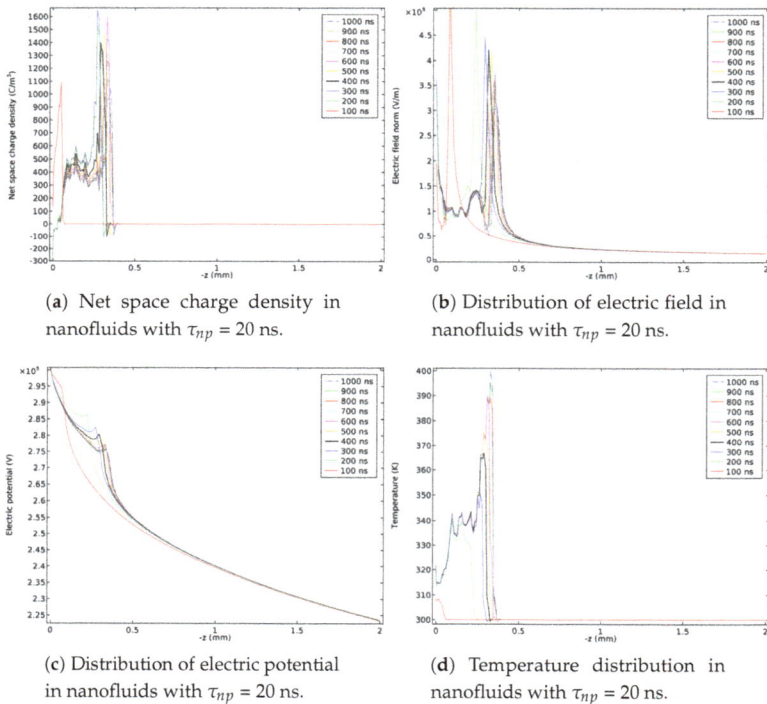

(**a**) Net space charge density in nanofluids with τ_{np} = 20 ns.

(**b**) Distribution of electric field in nanofluids with τ_{np} = 20 ns.

(**c**) Distribution of electric potential in nanofluids with τ_{np} = 20 ns.

(**d**) Temperature distribution in nanofluids with τ_{np} = 20 ns.

Figure 8. (**a**) Net space charge density, (**b**) distribution of electric field, (**c**) distribution of electric potential, and (**d**) temperature distribution along the z-axis from t = 0 to 1000 ns in intervals of 100 ns from the simulations in nanofluids with τ_{np} = 20 ns.

Table 4. Comparison between positive streamer propagation in mineral oil and mineral oil with nanoparticles.

U (kV)	Max Temperature (K)	Positive Streamer Length (z-axis; mm)	Velocity (km/s)	Time Constant (ns)
300	610	1.2	1.2	Mineral oil
300	425	0.53	0.707	50
300	400	0.415	0.593	20

4.3. Streamer in Base Fluid with Dielectric Barrier

When a dielectric solid is immersed with a perpendicular orientation to the electric field, the streamer is prevented from continuing to propagate; this forces the streamer to pass through or surround the solid, thereby delaying the electrical breakdown. Below, four time steps of simulations are shown for when a solid dielectric was immersed in the base fluid with three different relative permittivities.

For the three cases studied, $\varepsilon_r = 2.2$, $\varepsilon_r = 3.3$ and $\varepsilon_r = 4.4$, it was verified that the streamer impacted with the immersed solid as a result of its orientation (Figure 9a–c, respectively).

The greater the difference in the relative permittivity, the faster the streamer advanced over the surface of the dielectric solid. This effect is shown by comparison with the case of $\varepsilon_r = 2.2$ (Figure 9a), in which the streamer advanced more slowly than in the cases of $\varepsilon_r = 3.3$ (Figure 9b) and $\varepsilon_r = 4.4$ (Figure 9c).

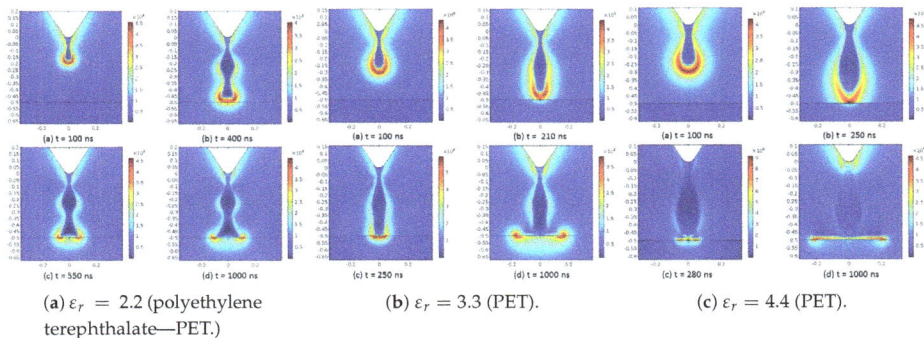

(a) $\varepsilon_r = 2.2$ (polyethylene terephthalate—PET.)

(b) $\varepsilon_r = 3.3$ (PET).

(c) $\varepsilon_r = 4.4$ (PET).

Figure 9. Positive streamer propagation. Evolution of the electric field (V/m) in mineral oil transformed with a horizontal solid barrier with (a) $\varepsilon_r = 2.2$ (polytetrafluoroethylene—PTFE), (b) $\varepsilon_r = 3.3$ (PTFE), and (c) $\varepsilon_r = 4.4$ (PTFE), applying 300 kV and with a rise time of 10 ns.

5. Challenges for Streamer Simulation

The main challenges for the simulation of the streamer in liquid dielectrics with finite-element software were as follows:

- Because of the scale of nanometers on which the simulations of the streamer were carried out, as well as the complex shapes of the geometry used, it was necessary to use a fine meshing (dividing the geometry into a high number of finite elements) to obtain results with sufficient precision. In addition to this, the number of linear and highly nonlinear equations introduced into the model had to be considered. Therefore, it was necessary to increase the number of cores used, for example, up to eight cores, and keep the frequencies for the processors above 3–3.5 GHz, working at the same time to be able to solve the model with acceptable times to carry out the simulations.
- Regarding the RAM memory, using a direct solver such as the multifrontal massively parallel sparse direct solver (MUMPS) reduced the use of RAM memory without requiring the use of a great resource of memory. This was possible because it stores part of the solution out-of-core, which means that part of the memory is stored on the hard disk. This behavior has the advantage of reducing the use of RAM memory, but it slows down the process, as it has to read data from the hard disk.
- The use of a graphics card was important at the time of the calculations and also at the time of visualizing the results. We note that it was not as critical a parameter as those previous. As an example, a Nvidia GeForce 940 with 4 Gb DDR3 had a good behavior for the simulations.

- Finally, to improve the performance and accuracy of the model, Equation (24) was used, which allowed us to reduce the oil rupture voltage to more realistic values:

$$G_I(|\vec{E}|) = \frac{e^2 n_0 \alpha |\vec{E}|}{h} exp(-\frac{\pi^2 m^* \alpha \; IP(\vec{E})^2}{eh^2 |\vec{E}|}) \tag{24}$$

where $IP(|\vec{E}|)$ is the electric field-dependent ionization potential function, whose value is obtained as follows:

$$IP(|\vec{E}|) = \Delta_0 - \gamma \sqrt{|\vec{E}|} \tag{25}$$

Here Δ_0 (eV) is the electrical potential needed for the molecular ionization and γ ($Jcm^{1/2}V^{-1/2}$) is the ionization potential function parameter. This theory is described by Jadidian in [9]. This model for dielectric liquids has been previously used to explain streamer propagation mechanisms for water and liquid dielectrics in [19–22]

- The difference between Equations (24) and (1) is that in the new equation, a continuous feedback of the value of the vector of the electric field is obtained, as a result of the new parameter $IP(|\vec{E}|)$. Therefore, an evolution that is more consistent with the reality of the field ionization ratio, $G_I(|\vec{E}|)$, is achieved.

6. Conclusions

The analysis of the dielectric and thermal behavior of the insulation system used in electrical assets such as the oil of power transformers is of interest for improving this kind of solid-liquid insulation system.

The work presented in this document allows us to have several models in finite-element software for analyzing the streamer behavior in several scenarios, such as in transformer oil, in a nanofluid and in transformer oil with solid dielectrics immersed. In accordance with the results obtained in this work, the following conclusions can be described:

- In mineral oil, a displacement of the potential and the electric field occurs as a result of the ionization of the oil (at field levels around 10^8 (V/m)). By ionizing more oil, a greater generation of charge carriers is obtained; this is shown in the figures on the net space charge density, generating a driving path where there is current conduction and therefore the advance of the streamer. This circulation of current causes an increase in temperature.
- When applying nanoparticles, the displacement of the positive streamer along the z-axis is reduced. In simulations with nanofluids and a time constant of 20 ns, there was a lower displacement of the field and potential in comparison with simulations made with a time constant of 50 ns; this was due to the fact that the nanoparticles were loaded faster for τ_{np} = 20 ns compared with τ_{np} = 50 ns—because of their influence in the streamer, this was accentuated when the time constant was reduced. As for the temperature, this produced something similar: temperatures were reduced, such that the deterioration of the oil was reduced.
- Regarding the barriers immersed inside the mineral oil, for the case of PTFE ($\varepsilon_r = 2.2$), as there was no difference in the relative permittivity between the dielectric solid and the oil, the dielectric solid did not attract the carriers of free charges by polarization; therefore, this propagation of the streamer on the dielectric solid was not due to the effects of polarization. However, as a result of the perpendicular orientation of the solid, the streamer, as it propagated, encountered the barrier and produced surface discharges on the dielectric solid that propagated along it. The PET ($\varepsilon_r = 3.3$) and Pressboard ($\varepsilon_r = 4.4$) cases were analogous, because their relative permittivities were higher than that of the oil in which they were submerged. Therefore, these dielectric solids attracted the carriers of free charges as a result of the polarization effect. On the other hand, because of the perpendicular orientation of the dielectric solid with respect to the main direction

of the electric field, this forced the streamer to meet with it. The greater the difference in relative permittivity compared to the oil, the greater the attraction by polarization and therefore the speed of propagation of the streamer on the surface of the dielectric solid. In addition to the greater difference in relative permittivity, the greater the force of attraction of the dielectric solid to the streamer, the greater the adherence of the streamer to the surface.

Acknowledgments: The authors wish to thank the Spanish Ministry of Economy and Competitiveness (project DPI2015-71219-C2-2-R) for supporting this work. Additionally, they would like to thank MDPI for promoting open access to knowledge, and even more, for providing discounts to reviewers, which helps to publish open access research such as this work.

Author Contributions: Juan Velasco and Ricardo Frascella implemented the models to carry out the streamer simulations and provided the results together with substantive suggestions and comments. Ricardo Albarracín guided the investigation and the writing of the article. Juan Carlos Burgos gave substantive suggestions and guidance for the research. Finally, Ming Dong, Ming Ren and Li Yang contributed in brainstorming for the analysis of the results.

Conflicts of Interest: The authors declare no conflict of interest.

References

1. Guerbas, F.; Adjaout, L.; Abada, A.; Rahal, D. Thermal aging effect on the properties of new and reclaimed transformer oil . In Proceedings of the IEEE International Conference on High Voltage Engineering and Application (ICHVE), Chengdu, China, 19–22 September 2016.

2. Yamada, H.; Sakamoto, S.; Nakao, Y. Studies of the breakdown process in dielectric liquids using high speed photography. *J. Electrostat.* **1979**, *7*, 155–168.

3. Lv, Y.; Ge, Y.; Li, C.; Wang, Q.; Zhou, Y.; Qi, B.; Yi, K.; Chen, X.; Yuan, J. Effect of TiO_2 nanoparticles on streamer propagation in transformer oil under lightning impulse voltage. *IEEE Trans. Dielectr. Electr. Insul.* **2016**, *23*, 2110–9878.

4. Lv, Y.; Ge, Y.; Du, Q.; Sun, Q.; Shan, B.; Huang, M.; Li, C.; Qi, B.; Yuan, J. Fractal Analysis of Positive Streamer Patterns in Transformer Oil-Based TiO_2 Nanofluid. *IEEE Trans. Plasma Sci.* **2017**, *45*, 1704–1709.

5. Oommen, T.V. Static electrification properties of transformer oil. *IIEEE Trans. Electr. Insul.* **1988**, *23*, 123–128.

6. Primo, V.A.; García, B.; Albarracín, R. Improvement of transformer liquid insulation using nanodielectric fluids; a review. *Electr. Insul. Mag.* **2018**, in press.

7. Hwang, J.G. Elucidating the Mechanisms Behind Pre-Breakdown Phenomena in Transformer Oil Systems. Ph.D. Thesis, Massachusetts Institute of Technology, Cambridge, MA, USA, 2010.

8. O'Sullivan, F.M. A Model for the Initiation and Propagation of Electrical Streamers in Transformer Oil and Transformer Oil Based Nanofluids. Ph.D. Thesis, Massachusetts Institute of Technology, Cambridge, MA, USA, 2007.

9. Jadidian, J. Charge Transport and Breakdown Physics in Liquid/Solid Insulation Systems. Ph.D. Thesis, Massachusetts Institute of Technology, Cambridge, MA, USA, 2013.

10. IEC. *Methods for the Determination of the Lightning Breakdown Voltage of Insulating Liquids*; IEC 60897; International Electrotechnical Commission: Geneva, Switzerland, 1987.

11. O'Sullivan, F.; Hwang, J.G.; Zahn, M.; Hjortstam, O.; Pettersson, L.; Liu, R.; Biller, P. A Model for the Initiation and Propagation of Positive Streamers in Transformer Oil. *IEEE Trans. Dielectr. Electr. Insul.* **2008**, in press.

12. Jadidian, J.; Zahn, M.; Lavesson, N.; Widlund, O.; Borg, K. Effects of Impulse Voltage Polarity, Peak Amplitude, and Rise Time on Streamers Initiated From a Needle Electrode in Transformer Oil. *IEEE Trans. Plasma Sci.* **2012**, *40*, 909–918.

13. Zener, C. A theory of the electrical breakdown of solid dielectrics. *Proc. R. Soc. Lond. Ser. A* **1934**, *145*, 523–529.

14. Girdinio, P.; Repetto, M.; Simkin, J. Finite Element Modelling of Charged Beams. *IEEE Trans. Magn.* **1994**, *30*, 2932–2935.

15. Prasath, R.T.A.R.; Roy, N.K.; Mahato, S.N. Mineral Oil Based High Permittivity $CaCu_3Ti_4O_{12}$ (CCTO) Nanofluids for Power Transformer Application. *IEEE Trans. Dielectr. Electr. Insul.* **2017**, *24*, 2344–2353.

16. Hwang, J.G.; Zahn, M.; O'Sullivan, F.M.; Pettersson, L.A.A.; Hjortstam, O.; Liu, R. Effects of nanoparticle charging on streamer development in transformer oil-based nanofluids. *IEEE Trans. Dielectr. Electr. Insul.* **2010**, *107*, 014310.

17. Beroual, A.; Zahn, M.; Badent, A.; Kist, I.C.; Schwabe, A.J.; Yamashita, H.; Yamazawa, C.; Daniis, M.; Chadband, W.G.; Torshi, Y. Propagation and structure of streamers in liquid dielectrics. *IEEE Electr. Insul. Mag.* **1998**, *14*, 6–17.

18. Segal, V.; Hjortsberg, A.; Rabinovich, A.; Nattrass, D.; Raj, K. AC (60 Hz) and impulse breakdown strength of a colloidal fluid based on transformer oil and magnetite nanoparticles. In Proceedings of the Conference Record of the 1998 IEEE International Symposium on Electrical Insulation (Cat. No.98CH36239), Arlington, VA, USA, 7–10 June 1998; Volume 2, pp. 619–622.

19. Devins, J.C.; Rzad, S.J.; Schwabe, R.J. Breakdown and prebreak-down phenomena in liquids. *J. Appl. Phys.* **1981**, *52*, 4531–4545.

20. Béroual, A.; Tobazéon, R. Prebreakdown phenomena in liquid dielectrics. *IEEE Trans. Electr. Insul.* **1986**, *E1-21*, 613–627.

21. Joshi, R.P.; Kolb, J.F.; Xiao, S.; Schoenbach, K.H. Aspects of plasma in water: Streamer physics and applications. *Plasma Process. Polym.* **2009**, *6*, 763–777.

22. Li, Y.; Zhu, M.X.; Mu, H.B.; Deng, J.B.; Zhang, G.J.; Jadidian, J.; Zahn, M.; Zhang, W.Z.; Li, Z.M. Transformer Oil Breakdown Dynamics Stressed by Fast Impulse Voltages: Experimental and Modeling Investigation. *IEEE Trans. Plasma Sci.* **2014**, *42*, 3004–3013.

energies

MDPI

Article

Impact of Low Molecular Weight Acids on Oil Impregnated Paper Insulation Degradation

Kakou D. Kouassi [1], Issouf Fofana [2,*], Ladji Cissé [1], Yazid Hadjadj [3], Kouba M. Lucia Yapi [2] and K. Ambroise Diby [1]

[1]   Ufr-SSMT Laboratory of Physics Condensed Matter and Technology, Université Félix Houphouet Boigny de Cocody-Abidjan, 22 BP 582 Abidjan 22, Côte d'Ivoire; lorellmongoua@gmail.com (K.D.K.); ladjic@hotmail.com (L.C.); dibyka@yahoo.fr (K.A.D.)
[2]   Research Chair on the Aging of Power Network Infrastructure (ViAHT), Université du Québec à Chicoutimi, Chicoutimi, QC G7H 2B1, Canada; KoubaMarieLucia1_Yapi@uqac.ca
[3]   Measurement Sciences and Standards, National Research Council Canada (NRC), Ottawa, ON K1A 0R6, Canada; yazidhad@yahoo.com
*   Correspondence: ifofana@uqac.ca; Tel.: +1-418-545-5011

Received: 30 April 2018; Accepted: 31 May 2018; Published: 6 June 2018

Abstract: Aging of a power transformer's insulation system produces carboxylic acids. These acids—acetic, formic and levulinic—are absorbed by the paper insulating material, thus accelerating the degradation of the whole insulation system. In this contribution, the effect of these acids on the aging of oil-impregnated paper insulation used in power transformer is reported. A laboratory aging experiment considering different concentrations of these three acids was performed to assess their effect on the insulation system's degradation. Each acid was individually mixed with virgin oil, and a mixture of acids was also blended with oil. The paper's degradation was assessed by the degree of polymerization (DPv). It was found that the DPv of paper aged with formic acid decreased much faster in comparison to the other acids.

Keywords: power transformer; oil-paper insulation; carboxylic acids; acetic acid; formic acid; levulinic acid; degree of polymerization

1. Introduction

Many power transformers installed around the world have exceeded their estimated design life, as the majority of them are approaching or exceeding the age of 25–30 years. However, these important machines are still in service well beyond their designed lifetime. With increasing age and exacerbated by the growing demand in electricity, there are potential risks of extremely high monetary losses due to unexpected failures and outages. Because most failures of aging power transformers are due to the degradation of its insulation system, it has become important to understand the aging of paper insulation in order to prolong the life of transformers, and to know the condition of their insulation, by means of suitable diagnostic tests. Increasing requirements for appropriate tools allowing diagnosing of power transformers non-destructively and reliably has promoted the development of modern complementary diagnostic techniques [1,2]. Research activities are still in progress worldwide to improve these techniques [1–6].

Transformer life/ageing is mainly related to the degradation of the oil paper insulation system. Heat, water, oxygen and aging products are severely contributing to transformer aging [7]. It is now well established that insulation aging is catalyzed by moisture, copper/copper alloys in aluminum windings and iron (which are primary transformer components), and oxygen [8]. The insulation aging may be subdivided in three processes:

(1)   Pyrolysis, promoted by heat;

(2)   Hydrolysis, by reactions with water; and

(3)   Oxidation by reactions with oxygen.

These processes are accelerated by heat, ageing by-products, dirt, vibrations/overload, and electrical stress/voltage waves, etc. [8].

Mineral oil used in transformers is mainly composed of three types of hydrocarbons. These hydrocarbons are paraffins, naphthenes and aromatics. They consist in 18 carbon atoms on average [9]. The insulating paper material used in power transformers consists largely of cellulose. Cellulose consists mainly as a polymer, formed from repeated monomers of glucose. As the oxidation level of in-service oil increases, polar compounds—particularly organic acids—form in the oil. During the aging of the oil/paper insulation system, acids are produced from chemical reactions. Carboxylic acids are generated from oxidation of the transformer oil or hydrolysis of its paper insulating material [10]. Because oil is more easily accessible, oil analyses are among the most used diagnostic techniques [1,2]. About 70% of the diagnostic information can be identified by oil analysis [1]. Among these techniques, the measurement of the total acid number (TAN) is frequently used to monitor the level of degradation of the insulating oil. Acid generation accelerates when water and oxygen are present. Acid in turn causes acid-hydrolysis. It is recommended that the oil be reclaimed when the acid number reaches 0.15 mg of potassium hydroxide (KOH)/g for power transformers with nominal voltages above 400 kV, 0.2 mg of KOH/g for 170 to 400 kV, and 0.3 mg of KOH/g for 72.5 to 170 kV [11].

Oil oxidation also produces low molecular weight acids which degrade the whole insulation. The TAN (consisting of large and low molecular weight acids) is the measure of acid concentration in a non-aqueous solution. The procedure involves determining the amount of potassium hydroxide (KOH) base required to neutralize the acid in one gram of an oil sample. The standard unit of measure is mg KOH/g. However, the oxidation of oil produces two types of acids: low molecular weight acids (LMA) and high molecular weight acids (HMA) (stearic acid [$CH_3(CH_2)16COOH$]) or naphthenic acids [R (C_5H_8) (CH_2) nCOOH] [12,13].

Three main reactions that could break this polymer chain are oxidation, pyrolysis and hydrolysis [14,15]. Acid hydrolysis degrades paper insulation by producing carboxylic acids such as formic and levulinic acids, which are low molecular weight acids (LMA). These carboxylic acids accelerate the degradation of paper in turn [12,16]. Actually, the "end-of-life" of the transformer = "end-of-life" of cellulose-based paper isolation [1,2].

High molecular weight acids are hardly soluble in oil and are not aggressive. Therefore, they do not intervene much in the degradation of the paper insulation [17,18]. On the contrary, low molecular weight acids (up to five carbon atoms) are more hydrophilic and have a greater influence on the degradation rate of paper throughout acid hydrolysis. The accumulation of acids in oil accelerates formation of insoluble deposits (x-waxes), which reduce the dielectric strength in turn. These deposits could also reduce heat transfer if they are deposited on paper or inside radiator pipes, increasing thermal degradation of paper insulation [19].

In recent years, research directed towards biodegradable liquids as alternatives to mineral oils has become popular. Significant differences exist between esters and typical mineral oils—especially the aniline point, flash point, interfacial tension, acidity, pour point, and viscosity. For example, it has been reported that the aging rate of the solid insulation in natural esters, even with a dark color and a very high acid value, is lower than that of the mineral oil impregnated solid insulation with a lower acidity value [20]. The TAN in oil is consequently not currently used to monitor aging of paper insulation, because the correlation between acidity and aging rate has not been fully determined [16]. Paper may contain 50% of acidity present in transformer insulation system. In addition, acid analyses indicate that hydrophilic acids were mainly concentrated in paper insulation. As a result, the majority of low molecular weight acids detected are located in paper. Also, the concentration of low molecular weight acid in paper is greater than in oil. Consequently, the polymerization degree (DPv) of paper insulation decreases [21].

To date, different aging indicators are being used to monitor transformer lifetime. Direct measurements to ensure that the paper continues to provide effective insulation for the transformer are performed through tensile strength or degree of polymerization, measured by the viscometric method (DPv). The DPv, which represents the average number of glucose units per cellulose chain, is directly correlated to the mechanical strength of paper [1]. To overcome the problems related to direct measurements, many studies were conducted to develop alternative indirect techniques for assessing the condition of solid insulation [1]. This article focuses specifically on the impact of low molecular weight acids on the degradation of oil-impregnated paper. The influences of three different LMAs—namely the acetic, formic and levulinic acids—on the DPv of Kraft paper impregnated with mineral oil are investigated. To simulate critical situations, a naphtenic-based mineral oil was used and samples with 0.2 mg KOH/g and 0.40 mg KOH/g of acid concentrations of the three acids, either individually or combined, were considered to simulate extreme degradations. A mixture of these three acids was also considered.

2. Background on Carboxylic Acids Formation in Transformer Oil

Mineral oil is mainly composed of three types of hydrocarbons, namely paraffins, naphthenes and aromatics, with an average number of carbon atoms equal to 18. When these hydrocarbon molecules (C_xH_y) and Oxygen molecules (O_2) react chemically, oil begins to deteriorate. This reaction forms the following compounds [8]:

- Acid deposits generate colloids. This colloidal suspension reduces the dielectric strength of the insulation system of the power transformer. Insoluble suspensions worsen the cooling effectiveness of the power transformer. The temperature increase degrades cellulose and acts as a catalyst for further oxidation.
- Water degrades cellulose reducing breakdown voltage and acts as a catalyst for oxidation. Water and acids have a multiplicative effect on the ageing rate [14].
- Gases (CO, CO_2, O_2, H_2, CH_4, C_2H_4, C_4H_6, C_6H_8) that are dissolved in the oil.
- Acids promote corrosion and also act as catalysts for the oxidation of oil.

All of these compounds, including carboxylic acids, are formed by a series of radical reactions [22]. Carboxylic acids formed in oil are classified into two groups: low molecular weight (LMA) and high molecular weight (HMA). Low molecular weight carboxylic acids (LMA) have an affinity with cellulose (paper) and water because of their ability to bond with the hydroxyl group ($-OH$). They react chemically to give a proton (H^+), as described by acid–base theory of Brønsted–Lowry [14,23].

3. Degradation of Cellulose

3.1. Swelling of Cellulose

Cellulose is a natural polymer. Macromolecules of cellulose are present in all plant species in the form of fibrils with very variable proportions. Wood (softwood or hardwood) contains between 40% and 50% of their weight as dry cellulose, whereas cotton fibers contain between 85% and 95% [24] of their weight. Cellulose is a polysaccharide of β-D glucans series [25]. Its repetitive pattern is cellobiose.

Cellulose has crystalline regions and amorphous regions. The hydrogen bonds are much more numerous in crystalline regions. Cellulose crystallinity degree varies from 40% to 50% for wood, 60% for cotton and more than 70% for certain seaweeds [26].

The biphasic structure of cellulose (both crystalline and amorphous) induces a two-step penetration of compound within cellulose. The first and easiest step is penetration of the amorphous zone of cellulose. It causes inter-crystalline swelling, which results in cleavage of β1-4glucosidic bond, hence depolymerization of the polymer chain. When compound penetrates crystalline regions, it induces intra-crystalline swelling. All compounds cannot reach the inner zone of crystal. Water, for example, only causes inter-crystalline swelling of cellulose fiber [27]. However, some compounds are able to pass through the crystallinity barrier and cause intra-crystalline swelling. Although penetrating cellulose deep within its structure, they are not able to dissolve it.

3.2. Acid Hydrolysis

The insulating paper consists of about 80% cellulose, 12% hemicelluloses, 8% lignin and certain mineral substances [24]. The aging of this insulating material has an impact on its electrical and mechanical strength. However, the main degradation factor of cellulose is acid hydrolysis, requiring water and acids [16]. Cellulose degradation will form more low molecular weight acids, which will remain absorbed in paper and accelerate the degradation of cellulose in turn [28,29].

Indeed, acid transfers a proton to water, forming hydronium, which then transfers the proton to cellulose, initiating chain splitting. According to the Bronsted–Lowry acid–base theory, the acid could transfer the proton directly to the cellulose, without an intermediate reaction with water [30]. However, water is a better acceptor of protons than cellulose because it is more polar. It is important to remember that there is a greater difference in electronegativity between oxygen and hydrogen in water than between oxygen and carbon in cellulose. The proton will share electrons with oxygen in the cellulose until the water causes the chain to split. Hydronium is particularly damaging because, after donating a proton to the cellulose, the remaining water molecule can then react with the cellulose. Note that a molecule of water is consumed during the chain scission, while the H+ ion is recovered. The H+ ion concentration remains unchanged. According to this reaction mechanism, the rate of degradation of the paper will depend on the concentration of water and H+ ions dissociated from the acids.

4. Materials and Methods

4.1. Oil and Paper Sample Preparation

The mineral oil used in this contribution is the Nytro Polaris GX. This oil was degassed and dehumidified for 48 h. Kraft paper samples had a surface area of 8×8 cm^2 were prepared. Paper and pressboard were dried in oven at 110 °C under vacuum for 24 h. The impregnation process was performed under vacuum. Usually, manufacturers establish a mass ratio of paper and oil according to the rating of the unit. In our investigations, the mass ratio between the cellulose insulation and oil used in these was 20:1. So, a 2000 mL of oil corresponding to a mass of 1812 g was used. This amount of oil corresponds to a mass of 90.60 g (paper + cardboard).

The impregnation of the paper samples with dried oil was performed under vacuum, in the same vacuum oven.

4.2. Oil Acidity Preparation

In addition to oil-free acid sample, two concentrations of each type of acid were prepared. Table 1 summarizes the chemical properties and the weight/volume of each acid added into 2 L of oil to obtain an acidity concentrations of 0.2 mg KOH/g and 0.4 mg KOH/g respectively.

Table 1. Summary of acids masses to be added in 2000 mL oil.

Acids	Purity	Molar Mass (g/mol)	Density (g/cm^3)	TAN (0.2 mg KOH/g)		TAN (0.4 mg KOH/g)	
				Weight (mg)	Volume (mL)	Weight (mg)	Volume (mL)
Acetic acid	0.997	60.05	1.0492	389.0364	0.3707	778.0728	0.7414
Formic acid	0.88	46.025	1.22	337.7568	0.2768	675.5136	0.5538
Levulinic acid	0.98	116.11	1.134	765.2076	0.6747	1530.4152	1.3494

Oil acidity preparation objective was to determine the amount of the acid when oil total acid number is known. This will allow preparing oil samples with the following acidity levels, which will be used for measurements:

- Paper with virgin oil,
- Paper with oil containing acetic acid of 0.20 mg KOH/g and 0.40 mg KOH/g,
- Paper with oil containing formic acid of 0.20 mg KOH/g and 0.40 mg KOH/g,
- Paper with oil containing levulinic acid of 0.20 mg KOH/g and of 0.40 mg KOH/g,
- Paper with oil containing the mixture of acid: (0.20 mg KOH/g of acetic acid) + (0.20 mg KOH/g of formic acid) + (0.20 mg KOH/g of levulinic acid).

The samples were aged in a convection oven at a temperature of 115 °C for a total time of 2000 h. Sampling was carried out every 500 h. A sample of each concentration was left at ambient temperature during 72 h to assess the effect of the acids on the insulation system. A Karl Fisher 831 KF coulometric titrator carried out the water content measurement according to ASTM (American Society for Testing Materials) D1533 standard.

The acidity measurement in transformer oil was performed according to ASTM D974 standard. An automated 862 Compact Titro-sampler system for titration the oil solution was used.

A capillary viscometer (Cannon-Fenske Routine) was used to measure the degree of polymerization (DPv), according to ASTM D4243 standard.

5. Results and Discussion

5.1. Assessing Moisture in Paper during Aging

Figure 1a–c shows the moisture changes in paper within aged oils. Three acids (formic, acetic and levulinic) individually mixed with various concentrations, a mixture of the three of them, and virgin oil (acid-free) were considered.

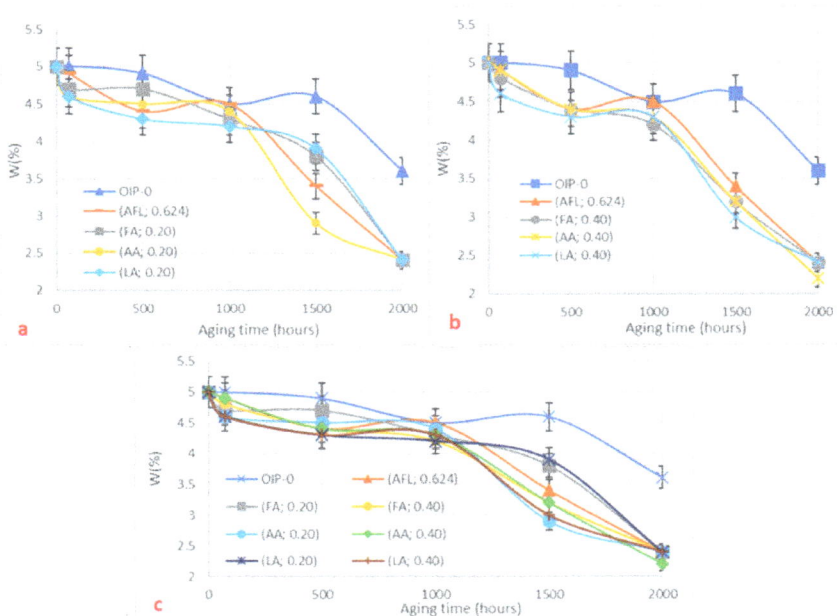

Figure 1. Moisture in paper as function of oil aging time within both acid contents: (**a**) 0.20 mg KOH/g and (**b**) 0.40 mg KOH/g for three different acids (formic, acetic and levulinic), acid mixture and also for acid-free oil. The results are summarized in Figure (**c**) for better comparison.

Figure 1a shows a decrease in moisture during aging time for all acids, with their various acid contents.

Water molecule absorption leads to polymer chain scission which in turn promotes paper insulation aging. So, polymer chain scission can be related to the reduction in moisture [23]. Figure 1b,c show that moisture in paper within virgin oil (HNAP) remained higher during aging. The amount of moisture decreased with aging time whatever the acid contents.

5.2. Total Acid Number (TAN) of Oil for Unaged Samples

Figure 2a–c presents acidity comparison before and after 72 h of exposure at room temperature. Three acids (formic, acetic and levulinic) individually mixed with various concentration, along with the mixture of them and virgin oil (acid-free) were considered. It is noted that after 72 h of aging at room temperature, acidity was lower in oils within formic acid for both concentrations (0.20 mg KOH/g and 0.40 mg KOH/g). Formic acid being of low molecular weight, and therefore highly soluble in oil. Accordingly, it could be absorbed readily by the insulating paper [12]. The same conclusion can be drawn for acetic acid with 0.20 mg KOH/g in oil compared to levulinic acid with same acid content. However, for higher concentrations of acid i.e., 0.40 mg KOH/g, the acidity of oil with acetic acid was higher than with levulinic acid.

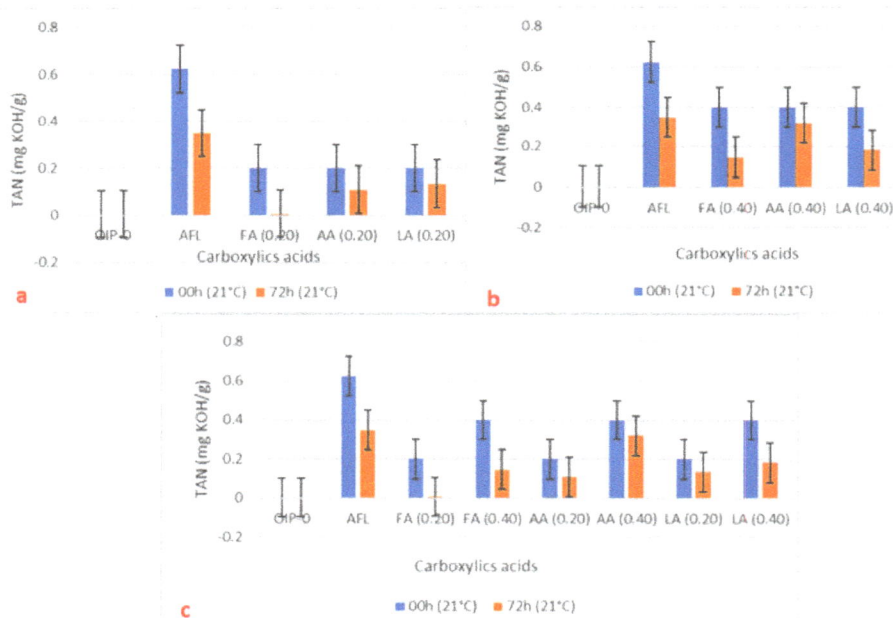

Figure 2. Acidity comparison before and after 72 h of exposure at room temperature: (**a**) 0.20 mg KOH/g and (**b**) 0.40 mg KOH/g for three different acids (formic, acetic and levulinic), acid mixture and also for acid-free oil. The results are summarized in Figure (**c**) for better comparison.

5.3. Degree of Polymerisation (DPv) before Thermal Aging

Figure 3a–c depicts the comparison of paper's DPv before and after 72 h of exposure at room temperature. Three acids (formic, acetic and levulinic) individually mixed with various concentration, along with the mixture of them and virgin oil (acid-free) were considered. These figures show that the degree of polymerization is lower for paper aged within oils containing formic acid after 72 h for both concentrations (0.20 mg KOH/g and 0.40 mg KOH/g). Formic acid is highly soluble in oil, and causes chain scission of the polymer chain of insulating paper [12]. On one hand, the DPv of paper with levulinic acid was lower than of paper exposed to acetic acid of 0.20 mg KOH/g. On other hand, the DPv of the paper with levulinic acid was higher than that of paper exposed to acetic acid of 0.40 mg KOH/g. For acid concentrations of 0.20 mg KOH/g, levulinic acid was more aggressive compared to acetic acid. This is in agreement with findings by Lungaard et al. [8,10,16,18] who reported that Kraft paper more readily absorbs formic acid than the acetic and levulinic acids.

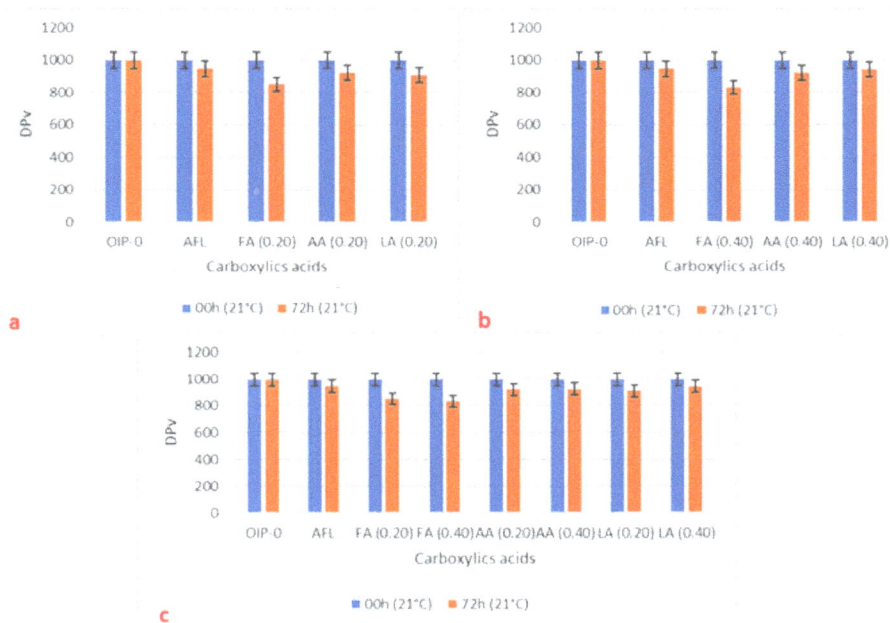

Figure 3. Degree of polymerization before and after 72 h exposure at room temperature: (**a**) 0.20 mg KOH/g and (**b**) 0.40 mg KOH/g for three different acids (formic, acetic and levulinic), acid mixture and also for acid-free oil. The results are summarized in Figure (**c**) for better comparison.

While the acidity of oil blended with mixed acids was reduced by almost half after 72 h of exposure at room temperature, the DPv of the paper samples remained almost unchanged. The acids mixture would delay the release of the H+ proton to cause degradation of paper cellulose.

5.4. Total Acid Number (TAN) of Oil for Aged Samples

Figure 4a–c show the evolution of the TAN during aging. In general, there was a decrease in the TAN for all acid concentrations up to 500 h aging. The absorption of carboxylic acids initially added in oil explains this drop in acidity [16]. This absorption may be related to the reduction in paper's polymerization degree (DPv) at this aging duration, which will be seen later. From 500 h up to 2000 h of aging, the TAN increases with aging time. This may be traced to the production of high molecular weight carboxylic acids that could not be absorbed by paper [17]. However, the results in Figure 4a indicate a low value of the TAN for 0.20 mg KOH/g of formic acid throughout the aging process. For oil containing 0.40 mg KOH/g of formic acid (Figure 4b), a low value of the TAN was observed after 72 h exposure at room temperature. However, during thermal aging, the TAN of oil with acetic acid was lower after 500 h (Figure 4b). The TAN of oil with formic and levulinic acids remained lower for

1500 and 2000 h aging, while the TAN of oil with mixed acid suddenly increased up close to 2000 h of aging. The increase in the TAN of oil with mixed acids during aging can certainly be traced to the generation of high molecular weight acids resulting from oil and paper degradation [6,8,16,18]. These findings are confirmed by partitioning studies reported by Lungaard et al. [8,18]. Indeed, at increasing temperature, more acids are moved to the oil due to an increased solubility with an activation energy much higher for formic acids.

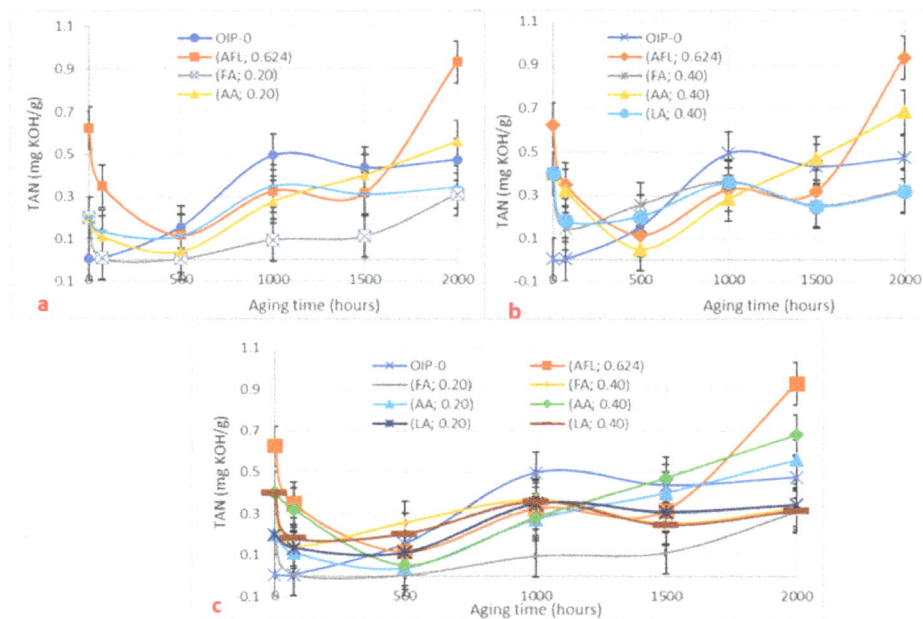

Figure 4. TAN (Total Acid Number) in oil as function of aging duration, for samples mixed with three different acids (formic, acetic and levulinic): (**a**) 0.20 mg KOH/g and (**b**) 0.40 mg KOH/g for three different acids (formic, acetic and levulinic), acid mixture and also for acid-free oil. The results are summarized in Figure (**c**) for better comparison.

5.5. The Degree of Polymerisation (DPv) during Thermal Aging

Figure 5a shows a comparison of the degree of polymerisation (DPv) of paper impregnated with oil containing formic acid as well as oil-free acids as function of aging duration. Generally speaking, the DPv of all samples aged in oils with both acid concentrations of 0.20 and 0.40 mg KOH/g decreased with aging, as expected. Some of the trend variations in the DPv values could be due to recombination of polymer bonds [31].

The higher amount of levulinic acid initially in oil would cause scission of bonds into the entire amorphous zone of paper up to 500 h, after which it would be impossible to cause bonds scission into the crystalline zone.

The DPv of paper aged in oil blended with formic acid decreased more rapidly than oils mixed with the other two acids (acetic and levulinic), regardless of the concentration. Figure 6a represents the DPv as function of aging for paper aged in oil blended with the mixture of three acids at a concentration of 0.624 mg KOH/g, compared to those aged in oil with formic acid of both acid concentrations (0.20 and 0.40 mg KOH/g). The DPv of paper aged in oil blended with formic acid for both acid concentrations decreased more rapidly than those aged with the acid mixture of 0.624 mg

KOH/g. From 500 h of aging, the DPv of paper aged in oil blended with the mixture of acids decreased more slowly, but remained relatively high. The bulkiness of these acids in oil could have certainly delay their dissolution.

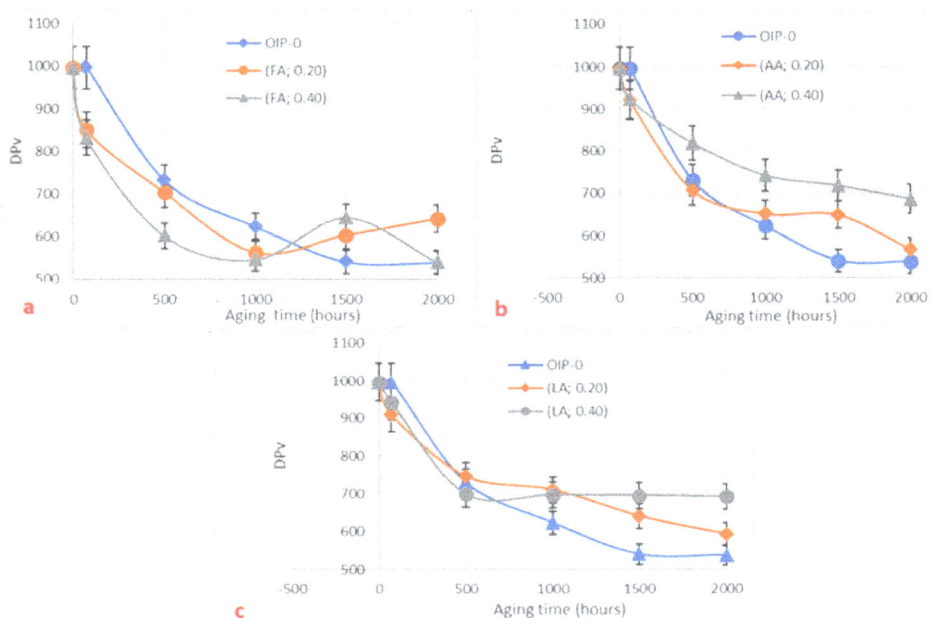

Figure 5. Fall in degree of polymerization (DPv) of paper during aging, for paper samples aged in oil mixed with three different acids (formic, acetic and levulinic) and virgin oil: (**a**) 0.20 mg KOH/g and (**b**) 0.40 mg KOH/g for three different acids (formic, acetic and levulinic), acid mixture and also for acid-free oil. The results are summarized in Figure (**c**) for better comparison.

In Figure 6, the DPv of the paper samples aged in different conditions is plotted against aging duration. The DPv of papers aged in acetic-acid-blended oil (0.20 mg KOH/g) and in the acid mixture at 0.624 mg KOH/g decreased more rapidly with aging compared to those aged in acetic-acid-blended oil at 0.40 mg KOH/g. However, the DPv within oil blended with acetic acid of 0.40 mg KOH/g remained higher. This result shows that the concentration of acetic acid is a limiting factor in the degree of paper polymerization. Large concentrations of acetic acid increases the DPv paper [32].

The DPv of paper aged in oil blended with levulinic acid (0.20 and 0.40 mg KOH/g) and oil blended with the acid mixture at 0.624 mg KOH/g decreased to 30% after 500 h aging. The values remained almost constant for oil blended with levulinic acid at 0.40 mg KOH/g, while those of paper aged in oil blended with levulinic acid at 0.20 mg KOH/g and the acid mixture were still decreasing. Reduction in the DPv might be traced to breaking of β1-4glucosidic bonds of amorphous cellulose regions [27]. High acid concentrations (0.40 mg KOH/g) of levulinic acid could cause all these bonds to break in this amorphous region from the first 500 h of aging. With no more amorphous regions [27], scission of bonds in the crystalline region of cellulose is virtually impossible, which could explain the stability of the DPv decrease from 500 h to 2000 h of aging.

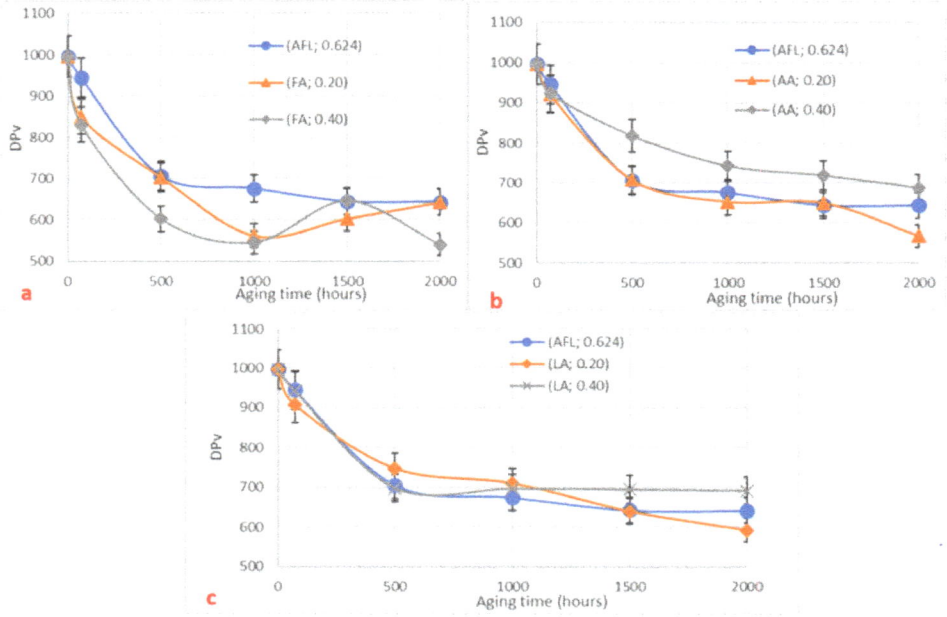

Figure 6. Degree of Polymerization (DPv) of paper as a function of aging in oil blended with acid mixture of 0.624 mg KOH/g, compared to samples aged in three different acids (formic, acetic and levulinic): (**a**) 0.20 mg KOH/g and (**b**) 0.40 mg KOH/g for three different acids (formic, acetic and levulinic), acid mixture and also for acid-free oil. The results are summarized in Figure (**c**) for better comparison.

Figure 7 represents the degree of polymerization as function of paper aging in oil within an acid mixture at 0.624 mg KOH/g compared to acid-free oil (neutral oil).

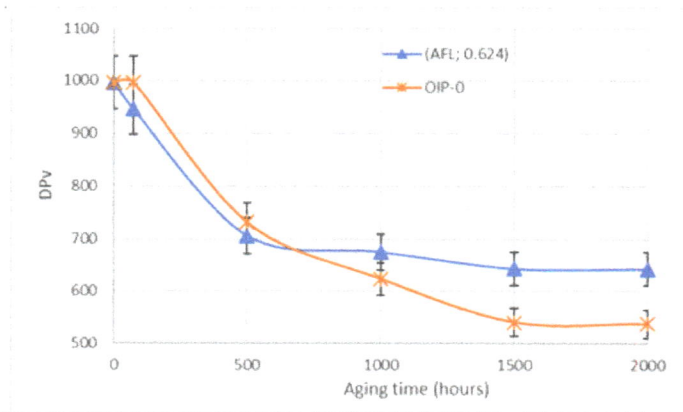

Figure 7. DPv of paper samples aged in oil blended with 0.624 mg KOH/g of the three acids compared to those aged in virgin oil as a function of aging duration.

Figure 7 shows that the DPv of paper aged in oil with the acid mixture was less than those aged in virgin oil up to 500 h of aging. From 500 h to 2000 h of aging, the DPv of paper aged in the presence of acid mixture became and remained higher than those aged in virgin oil.

By looking at Figure 1, it can be seen that up to 1000 h of aging, the moisture in insulating paper impregnated in both virgin oil (OIP) and the acid oil (AFL) remained high (>4%) which represents a critical value in transformers. At the same time, looking at Figure 4, it can be seen that during the first 500 h of aging, the acidity in the acid oil decreased rapidly from a value of about 0.6 mg KOH/g to about 0.1 mg KOH/g. This can be explained by the reaction of acid with the water molecules leading to a faster degradation of the insulating paper compared to the paper impregnated in virgin oil (OIP), where the acidity increased slightly due to oxidation phenomena during the first 500 h.

After this period, acidity increased gradually with aging in the two samples, but was more significant in the virgin oil than in the acid oil. However, in AFL, the moisture content of the paper (Figure 1) decreased brusquely to a value of 3.5% and then to 2.5%, at the same time the acidity increased sharply after 1500 h. This supports the idea that moisture and low molecular acids have a multiplicative effect [12,16,18]. This trade-off between acidity and humidity can lead to a compromise, which explains the "pseudo-variation" of the DPv (which decreased very slightly after 500 h of aging). In comparison, in OIP, humidity remained high throughout the aging period, at >3.5%, while the acidity increased gradually, which explains the decrease in DPv over time. Therefore, oil blended with acid mixture has a less harmful impact on the paper compared to virgin oil.

6. Conclusions

It is well established that the production of carboxylic acids during oil-impregnated paper aging results in a reduction in the degree of polymerization. In this paper, the effect of different carboxylic acids on the paper insulation degradation was studied. Because low molecular weight acids (LMA) are readily absorbed into the cellulose fibers, while high molecular weight acids (HMA) are not (because of their hydrophobic character), samples were aged in oil blended with different LMAs (formic, acetic and levulinic) at different concentrations. A mixture of these acids was also considered along with virgin oil to provide a baseline for comparison. From the obtained results, it was found that paper more readily absorbed formic acid than the acetic and levulinic acids. The aging rate of the paper was therefore more significant for samples aged in formic-acid-blended oil.

From these investigations, it may be concluded that present diagnostic techniques for monitoring insulation oil conditions based on the total acid number (which is used as reclamation criterion) does not provide a true picture of the transformer condition, since this procedure cannot distinguish between different acid types and their influences. Knowledge of the evolution of LMAs can be of great help to managers of transformer fleets, since the state of the dielectric paper determines the service life of a power transformer. New measuring techniques allowing discrimination of low molecular weight acids will be helpful in monitoring the condition of the solid insulation system.

Author Contributions: This work was done under the supervision of I.F., holder of the Research Chair on the Aging of Power Network Infrastructure (ViAHT), L.C. and K.A.D. (professors at Université Félix Houphouet Boigny) who designed this research and gave the whole guidance. K.D.K., Y.H. and K.M.L.Y. collected all the data, carried out calculations, result display and analysis and wrote the manuscript. The final draft of paper was thoroughly reviewed by I.F., L.C. and K.A.D. All authors read and approved the final manuscript.

Acknowledgments: This work was partially sponsored by the Natural Sciences and Engineering Research Council of Canada (NSERC) under Grant No. RGPIN-201504403. Thanks are also extended to the Government of Ivory Coast for supporting the research stay of Kakou in Canada.

Conflicts of Interest: The authors declare no conflicts of interest.

References

1. N'Cho, J.S.; Fofana, I.; Beroual, A. Review of Modern Physicochemical-based Diagnostic Techniques for Assessing Insulation condition in Aged Transformers. *Energies* **2016**, *9*, 367. [CrossRef]

2. Fofana, I.; Hadjadj, Y. Electrical-based Diagnostic Techniques for Assessing Insulation condition in Aged Transformers. *Energies* **2016**, *9*, 679. [CrossRef]

3. Peng, L.; Fu, Q.; Zhao, Y.; Qian, Y.; Chen, T.; Fan, S. A Non-Destructive Optical Method for the DP Measurement of Paper Insulation Based on the Free Fibers in Transformer Oil. *Energies* **2018**, *11*, 716. [CrossRef]

4. Godina, R.; Rodrigues, E.M.G.; Matias, J.C.O.; Catalão, J.P.S. Effect of Loads and Other Key Factors on Oil-Transformer Ageing: Sustainability Benefits and Challenges. *Energies* **2015**, *8*, 12147–12186. [CrossRef]

5. Cheng, L.; Yu, T. Dissolved Gas Analysis Principle-Based Intelligent Approaches to Fault Diagnosis and Decision Making for Large Oil-Immersed Power Transformers: A Survey. *Energies* **2018**, *11*, 913. [CrossRef]

6. Wang, X.; Tang, C.; Huang, B.; Hao, J.; Chen, G. Review of Research Progress on the Electrical Properties and Modification of Mineral Insulating Oils Used in Power Transformers. *Energies* **2018**, *11*, 487. [CrossRef]

7. Ward, B. *Application of Filtration System for On-Line Oil Reclamation, Degassing, and Dehydration*; Report No. 1002046; Electric Power Research Institute (EPRI): Palo Alto, CA, USA, 2003.

8. Fofana, I.; Sabau, J. The Service Reliability of Aging Power Transformers. In Proceedings of the Cigré Canada Conference, Calgary, AB, Canada, 9–11 December 2013.

9. Erdman, H.G. *Electrical Insulating Oils, STP 998*; ASTM International: West Conshohocken, PA, USA, 1988.

10. Xiang, Q.; Lee, Y.; Pettersson, P.O.; Torget, R.W. Heterogeneous aspects of acid hydrolysis of α-cellulose. In *Biotechnology for Fuels and Chemicals*; Number 1-3 Spring; Humana Press Inc.: New York, NY, USA, 2003; Volumes 107, pp. 505–514. ISSN 0273-2289.

11. *Mineral Insulating Oils in Electrical Equipment. Supervision and Maintenance Guidance*; IEC 60422 ed. 4.0; IEC International: Geneva, Switzerland, 2013.

12. Lelekakis, N.; Wijaya, J.; Martin, D.; Susa, D. The effect of acid accumulation in power-transformer oil on the aging rate of paper insulation. *IEEE Electr. Insul. Mag.* **2014**, *30*, 19–26. [CrossRef]

13. Martin, D.; Perkasa, C.; Lelekakis, N. Measuring paper water content of transformers: A new approach using cellulose isotherms in nonequilibrium conditions. *IEEE Trans. Power Deliv.* **2013**, *28*, 1433–1439. [CrossRef]

14. Lelekakis, N.; Martin, D.; Wijaya, J. Ageing rate of paper insulation used in power transformers Part 1: Oil/paper system with low oxygen concentration. *IEEE Trans. Dielectr. Electr. Insul.* **2012**, *19*, 1999–2008. [CrossRef]

15. Lelekakis, N.; Martin, D.; Wijaya, J. Ageing rate of paper insulation used in power transformers Part 2: Oil/paper system with medium and high oxygen concentration. *IEEE Trans. Dielectr. Electr. Insul.* **2012**, *19*, 2009–2018. [CrossRef]

16. Lundgaard, L.E.; Hansen, W.; Ingebrigtsen, S. Ageing of Mineral Oil impregnated Cellulose by Acid Catalysis. *IEEE Trans. Dielectr. Electr. Insul.* **2008**, *15*, 540–546. [CrossRef]

17. Ingebrigtsen, S.; Dahlund, M.; Hansen, W.; Linhjell, D.; Lundgaard, L.E. Solubility of carboxylic acids in paper (Kraft)-oil insulation systems. In Proceedings of the 2004 Annual Report Conference on Electrical Insulation and Dielectric Phenomena (CEIDP '04), Boulder, CO, USA, 17–20 October 2004; pp. 253–257.

18. Lundgaard, L.E.; Hansen, W.; Ingebrigtsen, S.; Linhjell, D.; Dahlund, M. Aging of Kraft paper by acid catalyzed hydrolysis. In Proceedings of the 2005 IEEE International Conference on Dielectric Liquids (ICDL 2005), Coimbra, Portugal, 26 June–1 July 2005; pp. 381–384.

19. CIGRE Task Force D1.01.10. *Ageing of Cellulose in Mineral-Oil Insulated Transformers*; CIGRE Technical Brochure No 323; CIGRE: Paris, France, 2007; 88p.

20. Bandara, K.; Ekanayake, C.; Saha, T.; Ma, H. Performance of Natural Ester as a Transformer Oil in Moisture-Rich Environments. *Energies* **2016**, *9*, 258. [CrossRef]

21. Ivanov, K.I.; Panfilova, E.S.; Kullkovskaya, T.N.; Zhakhovskaya, V.P.; Savinova, V.K.; Seminova, M.G. Effect of petroleum oil oxidation products on the ageing of paper insulation in transformers. *Zh. Prikl. Khim. (Leningrad)* **1974**, *47*, 2705–2711.

22. Prabhashankar, V.; Badkas, D.J. Mechanism of Oxidation of Transformers Oils. *J. Inst. Petr.* **1961**, *47*, 201–211.

23. Hill, G.C.; Holman, J.S. *Chemistry in Context*, 5th ed.; Nelson: Walton-onThames, UK, 2000.

24. Krassig, H.; Schurz, J. Cellulose. In *Ullmann's Encyclopedia of Industrial Chemistry*, 6th ed.; Wiley: Hoboken, NJ, USA, 2002.

25. Ruelle, J. Analyse de la Diversité du Bois de Tension de 3 Espèces D'angiospermes de Forêt Tropicale Humide de Guyane Française. Ph.D. Thesis, Université des Antilles et de la Guyane, Cayenne, France, 2006.

26. Lin, J.S.; Tang, M.Y.; Fellers, J.F. Fractal Analysis of cotton cellulose as characterized by small angle X-ray scattering. In *The Structures of Cellulose. ACS Symposium Series*; Atalla, R.H., Ed.; American Chemical Society: Washington, DC, USA, 1987; Volume 340, pp. 233–254.

27. Zeronian, S.H. Intercrystalline swelling of cellulose. In *Cellulose Chemistry and Its Applications*; Horwood: Chichester, UK, 1985; pp. 138–158.

28. Ojha, S.K.; Purkait, P.; Chakravorti, S. Evaluating the effects of lower molecular weight acids in oil-paper insulated transformer. In Proceedings of the 2017 3rd International Conference on Condition Assessment Techniques in Electrical Systems (CATCON), Rupnagar, India, 16–18 November 2017.

29. Wan, F.; Du, L.; Chen, W.; Wang, P.; Wang, J.; Shi, H. A Novel Method to Directly Analyze Dissolved Acetic Acid in Transformer Oil without Extraction Using Raman Spectroscopy. *Energies* **2017**, *10*, 967. [CrossRef]

30. Petrucci, R.H.; Harwood, W.S.; Herring, F.G. *General Chemistry*, 8th ed.; Prentice-Hall: Upper Saddle River, NJ, USA, 2002; p. 666.

31. Fellers, C.; Iversen, T.; Lindström, T.; Nilsson, T.; Rigdahl, M. *Ageing/Degradation of Paper—A Literature Survey*; Riksarkivet: Stockholm, Sweden, 1989; ISSN 0284-5636.

32. Kakou, D.; Cissé, L.; Fofana, I.; Hadjadj, Y.; Yapi, K.M.L.; Ambroise, D.K. Impact of Acidity on the Degradation of the Solid Insulation of Power Transformer. In Proceedings of the 20th International Symposium on High Voltage Engineering (ISH 2017), Buenos Aires, Argentina, 28 August–1 September 2017.

energies

MDPI

Article

Modeling the Insulation Paper Drying Process from Thermogravimetric Analyses

Amidou Betie [1,2], Fethi Meghnefi [1], Issouf Fofana [1,*] and Zie Yeo [2]

[1]   Research Chair on the Aging of Power Network Infrastructure (ViAHT), Université du Québec à Chicoutimi, Chicoutimi, QC G7H 2B1, Canada; betie.amidou1@uqac.ca or amidou.betie@inphb.ci (A.B.); fmeghnef@uqac.ca (F.M.)

[2]   Département Génie Électrique et Électronique, Institut National Polytechnique Houphouët Boigny, BP 1093 Yamoussoukro, Côte d'Ivoire; zie.yeo@inphb.ci or yeozie@gmail.com

*   Correspondence: ifofana@uqac.ca; Tel.: +1-418-545-5011; Fax: +1-418-545-5012

Received: 23 January 2018; Accepted: 22 February 2018; Published: 28 February 2018

Abstract: It is now well-established that moisture in the oil paper insulation used in power and instrument transformers significantly reduces the transformers' lifetimes, and can eventually lead to premature failure. This moisture should, therefore, always be removed, not only during production but also after repairs. At the final stage of manufacturing, the drying process should be carried out to remove water and air vacuoles contained in the cellulose-based paper before impregnation. Successful drying helps increase the residual life of transformers, because the presence of moisture and air vacuoles accelerates the aging/degradation process of the oil paper insulation. Proper estimation of residual moisture before impregnation and the determination of the time required for drying play key roles in the time-consuming process of drying. In this paper, the disadvantages of inadequate drying are addressed, followed by a mathematical approach to model the paper drying process. A mathematical model describing the kinetics of drying according to temperature, initial moisture, paper weight, final moisture, and extraction rate is proposed. This model also estimated the amount of moisture removed at the end of the drying process.

Keywords: power transformers; dielectric dissipation factor; oil impregnated paper insulation; moisture; drying process; drying curves; diffusion coefficient; heating process

1. Introduction

Power transformers are vital for the production, transmission, and distribution of electrical energy [1,2]. They are used to adapt the voltage level to the needs of users. In addition to playing this role, they are also the most expensive equipment found in transformation stations. Indeed, a transformer represents about 60% of the total cost of a power station, so the costs associated with repair or replacement, when a failure occurs, can often reach millions of dollars [3]. The proper operation and lifetime of power transformers are strongly linked to the state of the oil paper insulation used for their windings.

The presence of moisture in the transformer leads to the degradation of the insulation system, and accelerates the aging process of the paper and the oil. This degradation increases the probability of unexpected failures. Indeed, due to the highly hygroscopic nature of cellulose, paper insulation can contain up to 8% moisture at the end of the transformer manufacturing process [4]. Elimination or reduction of moisture has become a challenge for power transformer manufacturers, who want to offer their customers high quality products with acceptable residual moisture. To attain this result, transformer manufacturers dry their paper under a vacuum (<1 mbar) at temperatures between 85 °C to 130 °C, for a period ranging from several days to two or three weeks before impregnation, depending upon the size of the transformers [5–8]. This operation aims to remove water vapor and air in the paper interstices before the impregnation process [9]. Drying allows for the reduction of the water content in the solid insulation to a level of 0.25% or less.

The drying of the active parts before placing them in the tank is a key step in the construction of transformers because its success affects the useful life of the transformer. Indeed, if the drying process is done properly, transformer failures are greatly reduced. However, improper drying may present the following hazards:

- the initiation of partial discharges may cause gas bubbles, and thereby the release of hydrogen (the creation of partial discharges becomes significant when the moisture level is above 3%) [10,11];
- increased dielectric dissipation factor (tan δ) and decreased breakdown voltage (U$_d$), which would place the voltage outside acceptable limits [12].

Improper drying can also lead to the degradation and accelerated aging of the insulation system (oil paper). The presence of water leads to an increase in the de-polymerization speed of the paper [12], thus reducing the transformer's lifetime [4,13,14]. The new transformer paper has the degree of polymerization (DP) of 1200 and end-of-life DP is found to be around 200. The relative de-polymerization speed is computed from the actual measured DP, using the following formula:

$$\frac{1200 - DP}{1200} \tag{1}$$

To avoid failure during the dielectric testing certification after manufacture, the factors of duration, drying temperature, and estimation of residual moisture are all important, and their control ensures acceptable residual moisture. At present, the length of the drying process is still too often based on manufacturer experiences and empirical procedures.

During the drying process, the water contained in the paper is released by heating the windings. This water is collected and, depending to the extraction rate, it may be decided to stop the drying process. The major drawback of this practice is that the heating may be stopped while there is still a significant amount of water in the paper. The determination of the required drying time plays a very important role in this process. Moreover, successful drying contributes to an increase in the residual life of the cellulose insulation.

Since the drying process—for example, in a vacuum oven–is extremely time-consuming, anything which can reduce the drying time offers a major benefit. For example, a device allowing for the monitoring the drying process of oil paper insulation in real time using a dielectric response (a tan δ measurement over a wide frequency range) has recently been introduced on the market [15].

It is in this context that the cost-effective analytical approach presented in this paper is proposed. A mathematical model describing the kinetics of drying according to temperature, initial moisture, paper weight, final moisture, and extraction rate is proposed. This model will allow the following:

- estimating the moisture removed at the end of drying, depending on initial moisture and drying temperature
- determining the time needed to obtain a better drying result for the paper.

In further research the authors are going to consider in the mathematical model other parameters which plays an important role in drying process in the mathematical model, like the pressure, mass, thickness, and density of cellulose materials.

2. Background

Drying can be considered to be a mass transfer, because it consists of extracting an amount of water contained in the paper [9,16]. For example, a 150 MVA–400 kV transformer containing about 80,000 L of oil and 7 tons of paper can also contain 490 L of water, if the moisture content of the paper is 7% [5,17]. Insulation paper drying will consist, therefore, of extracting the water present in the paper by heating.

The first mathematical model of the drying process was proposed by Lewis in 1920, to be followed by those of Sherwood (1929) and Newman (1931) [18].

The work of these previous studies showed that in the case of moisture removal (drying) in one dimension, the diffusion equation can be written using Fick's second law (Equation (2)). The validity of this equation implies that the moisture transfer direction is normal for a sheet of paper of thickness equal to 2a.

$$\frac{\partial M}{\partial t} = \frac{\partial}{\partial x}\left(D\frac{\partial M}{\partial x}\right)(0 < x < a,\ t > 0) \tag{2}$$

where M is the moisture content (in %), D is diffusion coefficient (in m^2/s), x is the humidity transfer direction, and t is the time (in s).

The solution of the diffusion Equation (2) was given by Sherwood (1929) and followed by Crank (1975) [19,20], for a thin layer of paper and an assumed constant diffusion coefficient, as follows:

$$MR = \frac{M - M_e}{M_0 - M_e} = \frac{8}{\pi^2}\sum_{n=0}^{\infty}\frac{1}{(2n+1)^2}exp\left(-\frac{(2n+1)^2\pi^2}{4}\frac{D}{L^2}t\right) \tag{3}$$

where MR is the humidity ratio, M_e is the equilibrium moisture content (in %), M_0 is the initial moisture (in %), M is the moisture at time t (in %), D is the diffusion coefficient (m^2/s), L is the half thickness of the layer (m), t is the drying time (s), and n represents the number of terms considered.

According to Ibrahim et al., the values of M_e are relatively low compared to M or M_0, so the error involved in the simplification is negligible. Hence, the moisture ratio can be expressed as follows [16]:

$$MR = \frac{M - M_e}{M_0 - M_e} = \frac{M}{M_0} \tag{4}$$

In addition, for longer drying periods and moisture less than 0.6%, Equation (3) can be simplified by limiting it to the first term in the series only, without unduly affecting the prediction accuracy. Thus, it can be written that [20]:

$$MR = \frac{M}{M_0} = \frac{8}{\pi^2}exp\left(-\frac{\pi^2}{4L^2}\cdot D\cdot t\right) \tag{5}$$

or that:

$$M = \frac{8}{\pi^2}exp\left(-\frac{\pi^2}{4L^2}\cdot D\cdot t\right)\cdot M_0 \tag{6}$$

From investigations mainly focused on the determination of the diffusion coefficient of the impregnated and non-impregnated paper and pressboard used in power transformers, Garcia et al. proposed the following equation for determining the diffusion coefficient [21,22]:

$$D = 61.627 \cdot l^{-5.431} \cdot e^{\left(0.2\cdot C_m - \frac{9380.7\cdot l^{-0.334}}{T_k + 273}\right)} \tag{7}$$

where D is the diffusion coefficient (in m^2/s), C_m represents the moisture content (in %), l is the insulation thickness (mm), and T_k is the drying temperature (in °C).

Empirical relationships have been established and are used as alternative solutions to Equation (7). These equations also facilitate the study of thin layer drying systems. The most commonly used of these equations are presented in Table 1. In order to select the best-suited empirical equation to the system under study, it is necessary to determine which one provides the best correlation with the experimental curve.

Table 1. Mathematical models applied to drying curves [23].

Model Name	Model Expression
Newton	$MR = exp\,(-kt)$
Page	$MR = exp\,(-kt^n)$
Modified Page 1	$MR = exp\,(-(kt)^n)$
Modified Page 2	$MR = exp\,((-kt)^n)$
Henderson and Pabis	$MR = a\,exp(-kt)$
Logarithmic	$MR = a\,exp\,(-kt) + c$
Two-term	$MR = a\,exp(-k0\,t) + b\,exp\,(-k1\,t)$
Two-term exponential	$MR = a\,exp(-kt) + (1-a\,)exp(-k\,a\,t)$
Wang and Singh	$MR = 1 + at + bt^2$
Diffusion approach	$MR = a\,exp(-k\,t) + (1-a\,)exp(-k\,b\,t)$
Modified Henderson and Pabis	$MR = a\,exp\,(-kt) + b\,exp\,(-gt) + c\,exp\,(-ht)$
Verma et al.	$MR = a\,exp(-kt) + (1-a\,)exp(-g\,t)$
Midilli-Kucuk	$MR = a\,exp\,(-kt^n) + bt$

3. Experimental Setup

The drying experiments were performed on 6 mm × 6 mm paper samples of 0.05 mm thickness, with high density (1.02 g/cm^3) and thermally upgraded, electrical grade creped 12 HCC, manufactured by Weidmann Electrical Technology Inc. (St. Johnsbury, VT, USA).

Prior to the drying process, the initial moisture content and degree of polymerization of the paper samples were determined. The drying was carried out in an oven. Different drying environments were simulated by varying vacuum pressure and temperature. The impact of the air/oxygen inlet on the quality of the paper was also assessed. The investigations were performed at different temperatures: 115, 130, and 150 °C. At specific time intervals (60, 120, 180, 240, 300, and 360 hours), some paper samples were removed to determine their degree of polymerization.

In order to measure and draw the drying curves of paper samples as a function of temperature, weight, and initial moisture, the thermogravimetric analysis technique (TGA), which consists of measuring the weight change of a sample as a function of time for a given temperature or temperature profile, was used [24]. As the moisture contained in the paper is progressively removed by heating, the weight begins to decrease, until it reaches an equilibrium position, which generally corresponds to the state when almost the entire amount of water in the paper is removed by evaporation. The moisture at each time step is obtained by the following formula [24]:

$$Moisture(\%) = \frac{W - D}{W}100\% \qquad (8)$$

where W is the initial weight (weight of wet paper), and D represents the paper weight measured at each instant (dried paper weight).

The measurement is stopped, and the value of the moisture recorded, when the variation of moisture according to time is less than or equal to the drift value (% per min) set on the device.

The measurement was carried out with the MF-50 Moisture Analyzer [24].

This thermogravimetric analysis technique (TGA) device allows for drawing distinct curves representing the moisture extracted, according to the variation of the drying temperature, as well as the weight and thickness of the paper to be dried.

The analysis of the resulting curves allows us to establish a mathematical model characterizing the drying process. However, it is important first to select among the empirical equation models in Table 1, the one which is best suited to the system under study.

This equation is to be used to model the drying time of the paper depending on various parameters, such as paper weight, drying temperature, extraction rate, initial moisture, etc.

4. Experimental Procedures

Various tests were conducted in order to determine the impact of different parameters on the drying process.

4.1. Weight Impact on the Drying Process

To assess the impact of the amount of water extracted from the paper samples with different weights were considered for a given temperature. This test makes it possible to predetermine the drying time necessary, depending on the paper's weight, and therefore transformer size, initial moisture, and drying temperature, allowing us to determine a mathematical model of the drying process.

The measurement results displayed in Figure 1 indicate the progression of moisture percentage extracted over time, depending on the weight of the paper to dry. The drying apparatus stops the measuring process and displays the final moisture when the extraction rate reaches a threshold of 0.02% per min.

The measurement results show that the drying time increases according to the weight of the paper to be dried. Indeed, moisture refers to the water (liquid or vapor) content in the air or another substance, and is defined as the ratio between the water weight and the weight of the substance, as follows:

$$H(\%) = \left(\frac{water\ weight}{Total\ weight} \right) 100 \tag{9}$$

so that:

$$water\ weight = H(\%) \times Total\ weight$$

This relationship indicates that for given moisture level, the weight of water in a substance is proportional to the weight of this substance. In this case, the relationship reflects the increase in extracted water that explains the increase in the drying time.

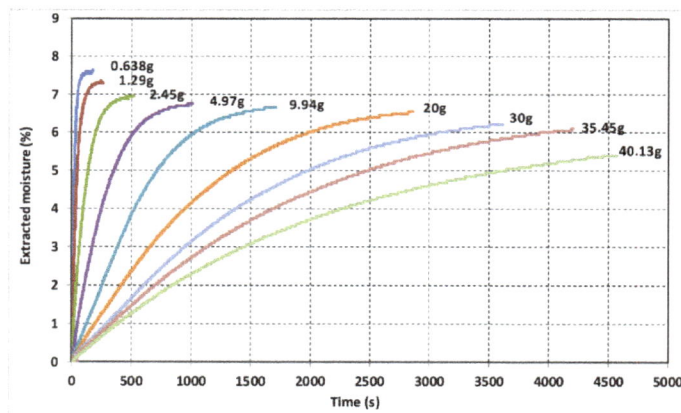

Figure 1. Influence of weight on drying time (130 °C).

4.2. Temperature Impact on the Drying Process

Given the large size of power transformers, some of which may contain several tons of paper, drying requires high temperatures (up to 125 °C and more). The objective of this experiment is to estimate how drying time varies with temperature. The results are shown in Figure 2.

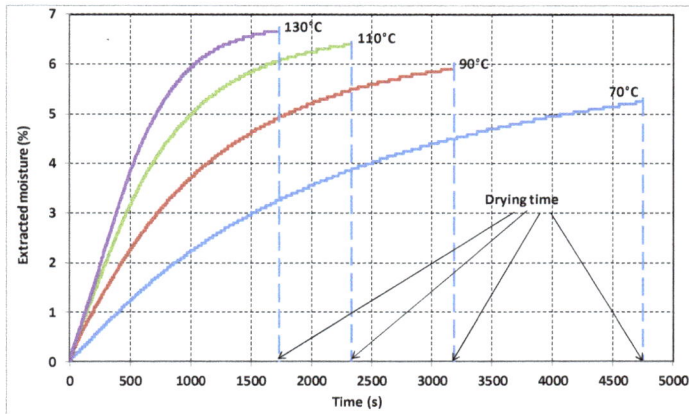

Figure 2. Influence of temperature on the drying time of a weight of 10 g.

Analysis of the results indicates that drying time decreases as temperature rises. According to Equation (7), the variation of moisture depends on diffusion coefficient D in (m^2/s); this coefficient increases with temperature, causing a reduction in the drying time.

5. Results and Discussion

5.1. Mathematical Model of the Drying-Out Process of Insulation Paper

Manufacturers monitor insulation dryness during processing, usually by measuring some parameter (such as insulation resistance or power factor) that is directly dependent on moisture content. Since there are no absolute values for these parameters applicable to all transformers, readings are usually plotted graphically, and the drying out process is considered completed when a levelling-out of the power factor and a sharp rise in insulation resistance is observed [5].

The first step in modeling the drying process is to determine the mathematical equation model that best fits the drying curves obtained. Since the percentage of extracted moisture is monitored, and not the percentage of residual moisture, as expressed by the equations shown in Table 1, it is therefore necessary to adapt these equations.

To do this, the Newton model (first equation in Table 1) seems most appropriate to fit the growth curves (extracted moisture), rather than the decreasing curves (residual moisture). The model used in this case is as follows:

$$H_{ext} = H_0 \left(1 - e^{-\frac{t}{\tau}}\right) \tag{10}$$

with H_{ext} as the extracted moisture (in %), H_0 as the final moisture (in %), τ as the time constant (in s), and t as the drying duration.

This equation determines the remaining residual moisture in the paper at the end of drying.

5.2. Residual Moisture Estimation

The approximation of the drying curves by Equation (10) is performed by optimizing the quadratic error between the measured drying curves and the curves calculated by Equation (9). The results of this optimization will allow calculation of the final moisture H_0 and time constant τ. The drying phenomenon can be modeled as a transient evolution, as the time constant of Equation (10) characterizes the time at the end of which equilibrium is reached—this is to say, when the extraction rate reaches a threshold of 0.02% per min. This optimization is achieved by using the MATLAB (R2013b) function algorithm "fminsearch", which allows one to find the minimum of an unconstrained multivariable function using a derivative-free method. The multi-variables to be determined are the time constant and the final moisture percentage, and the function is Equation (9). The extraction rate (% per min) is determined by the derivative of Equation (10). The estimated moisture is calculated when the extraction rate is equal to 0.02% per min (the threshold for which the drying apparatus has been set to stop drying and display the result of moisture).

Table 2 shows time constant τ and moisture H_0, calculated from the drying curves shown in Figure 1. The extraction rate (% per min) is determined by the derivative of Equation (10). The estimated moisture is calculated when the extraction rate is equal to 0.02% per min (the threshold for which the drying apparatus has been set to stop drying and display the result of moisture).

Table 2. Calculated results of the drying time constant and calculated moisture.

Weight (g)	Measured Moisture (%)	Calculated Moisture (%)	Time Constant (s)
Drying Temperature: 70 °C			
0.658	6.45	6.51	121.97
1.258	6.10	6.03	166.97
2.558	5.75	5.75	506.91
5.288	5.40	5.58	1009.31
10.286	5.25	5.84	2142.34
20.808	4.70	5.67	3323.47
25.642	4.45	5.82	4123.22
30.634	4.05	5.43	4461.33
35.096	4.05	5.63	4931.61
Drying Temperature: 90 °C			
0.63	6.55	6.49	40.29
1.282	6.20	6.16	86.63
2.53	6.15	6.20	228.24
4.912	6.10	6.37	472.79
9.892	5.90	6.29	1136.10
20.348	5.45	6.05	1981.50
31.22	5.05	6.07	2832.05
35.856	4.70	6.06	3713.64
Drying Temperature: 110 °C			
0.658	7.15	7.11	20.98
1.268	7.00	6.97	49.64
2.514	6.50	6.79	148.23
5.168	6.60	6.85	347.95
10	6.40	6.74	739.33
20.11	6.00	6.90	1908.63
30.11	7.10	8.09	2446.64
35.432	6.95	8.19	3175.70
Drying Temperature: 130 °C			
0.638	7.60	7.58	17.96
1.29	7.30	7.31	37.22
2.454	6.95	7.09	110.06
4.978	6.75	7.03	277.48
9.942	6.65	7.25	589.34
20.094	6.50	7.24	1131.06
30.042	6.20	7.07	1601.35
35.45	6.10	6.92	1927.77
40.13	5.40	6.16	2156.22

The approximation of the drying curves by Equation (10) not only allowed us to determine different time constants, but also to estimate the remaining residual moisture in the paper after the drying cycle.

Figure 3 shows an example of a calculated drying curve using Equation (10). For 0.658 g of paper, the moisture analyzer indicates 6.45 %, while the calculated moisture is 6.51%. It can be seen from the estimated curve that the real moisture contained in the paper can be calculated and that the remaining residual moisture in the paper can be estimated after the drying cycle.

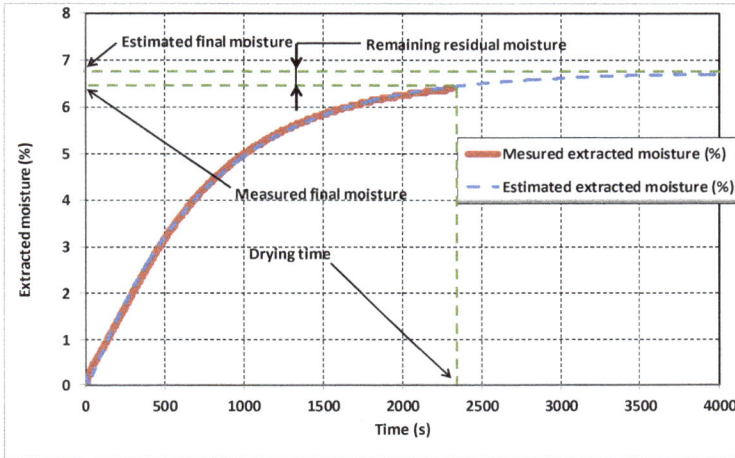

Figure 3. Principle of the estimation of the real and residual moisture.

Indeed, the apparatus used stops the drying when the moisture extraction rate, expressed in percentage per minute (% per min), reaches a previously set threshold. Thus, as can be seen from the various curves at the beginning of drying, the extracted moisture amount per time unit is very large, which explains its extreme slope. At the end of drying, the extracted moisture per time unit (the drying curve slope) becomes increasingly weak, eventually reaching the adjusted threshold value when drying is stopped. In this case, and according to the adjusted threshold of the moisture extraction rate, the heating may be stopped, even though there is still a certain amount of residual water in the paper.

The estimation process of the drying state of the transformers at the factory follows the same principle. In fact, during the drying process, the amount of removed water per hour is measured in grams. Then, depending on the transformer paper weight to dry, the extraction rate, expressed in grams of water per ton of paper per hour $(g/T/h)$, is calculated. When this extraction rate reaches a specific value, the drying operation is stopped.

Table 2 shows that in terms of the mass of paper to be dried and the drying temperature used, a significant quantity of water can still remain in the paper. This fact explains the difference between the extracted moisture value measured by the device and the value of estimated real moisture.

In addition, from the results of this analysis, it can be seen that the residual moisture in the paper after the drying process is higher if the drying time constant is large, as shown in Figure 4, where residual moisture is normalized to the initial moisture. It may be noted that in this case, at the end of a drying process, more than 25% of the initial moisture can remain, if the drying is done with a close time constant of 5000 s.

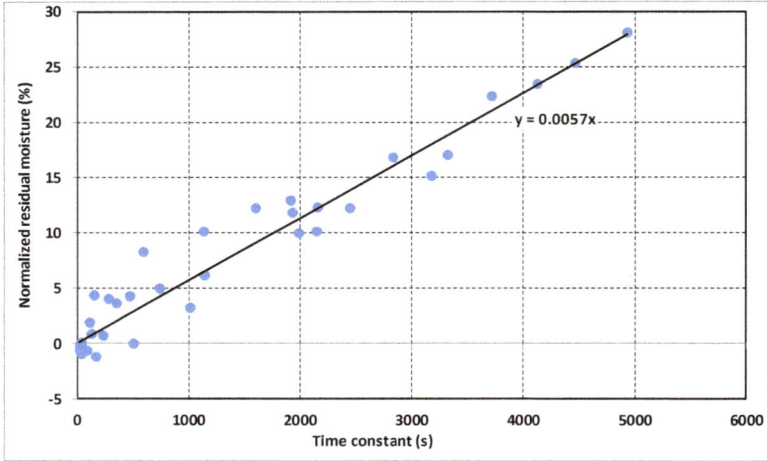

Figure 4. Impact of drying time constant on residual moisture.

This increase in residual moisture, in accordance with the increase of the time constant, is due mainly to the use of a fixed extraction rate threshold to stop the drying process, no matter the weight of the paper transformer or the drying temperature. These last two parameters directly influence the time constant, as already emphasized in this analysis. However, the use of an extraction rate threshold that takes into consideration the time constant (which depends on the paper weight and the temperature) can reduce the residual moisture in the paper. Therefore, the drying time in this case will increase, thereby contributing to the exposure of the paper to the drying temperature for longer periods.

In practice, it would be highly desirable in the industry to model the drying curves, in order to estimate the moisture removed at the end of drying, based upon initial moisture, weight of the paper, and drying temperature. In addition to the estimated moisture value, it would be possible to determine the time needed for the best method of drying paper.

5.3. Modeling the Drying Process Depending on Temperature and Weight

The study of the parameters influencing paper-drying kinetics in the transformers has resulted in the mathematical model represented by Equation (10). This model, which is perfectly adapted to the drying process, can estimate the final moisture H_0 and time constant τ.

From this equation, it is possible to calculate the duration of drying time, depending on paper weight and drying temperature.

In this case, the results obtained by estimating the extracted moisture depending on temperature and paper weight to be dried (Table 2) will be used.

Figure 5, obtained from the data in Table 2, shows the estimated time constant, depending on the temperature and the paper weight.

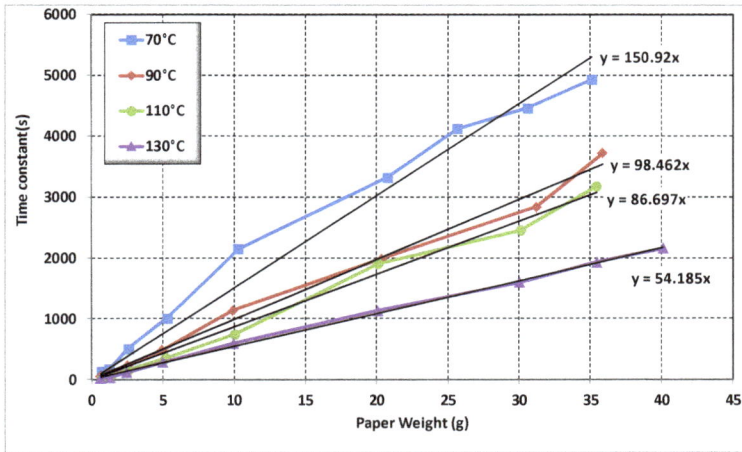

Figure 5. The time constant evolution, depending on the paper weight and drying temperature.

The analysis of the curves shows that time constant τ is directly proportional to the weight of the paper to be dried, as well as the drying temperature. The evolution of time constant τ can be estimated by linear curves. Therefore, the equation that relates the drying time constant τ and the paper weight M can be written as follows:

$$\tau = P \cdot M \tag{11}$$

where P is the slope of the linear approximation curves of the time constant, depending on the weight of the paper (M). The P slope depends on the drying temperature, and its evolution is presented in Figure 6.

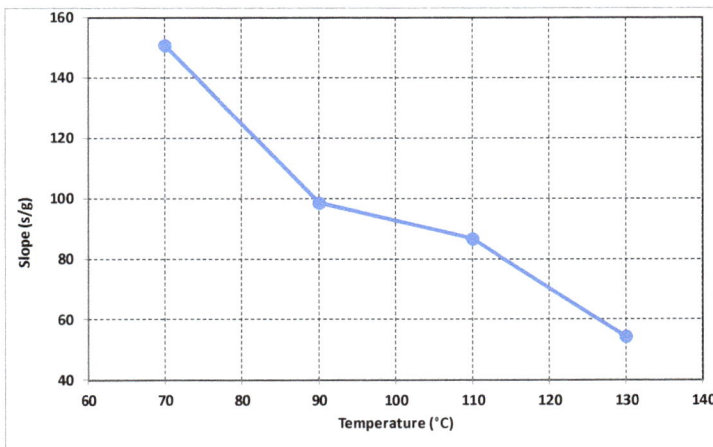

Figure 6. The slope of the linear approximation curves as a function of temperature.

The approximation function of the slope P as a function of temperature can be expressed as:

$$P = K_1 \cdot e^{-\frac{T}{K_2}} \tag{12}$$

where K_1 and K_2 are two coefficients, to be determined by minimizing the square root error between the slopes of Figure 8 and those calculated by Equation (10). Figure 7 displays the approximation function of the slopes.

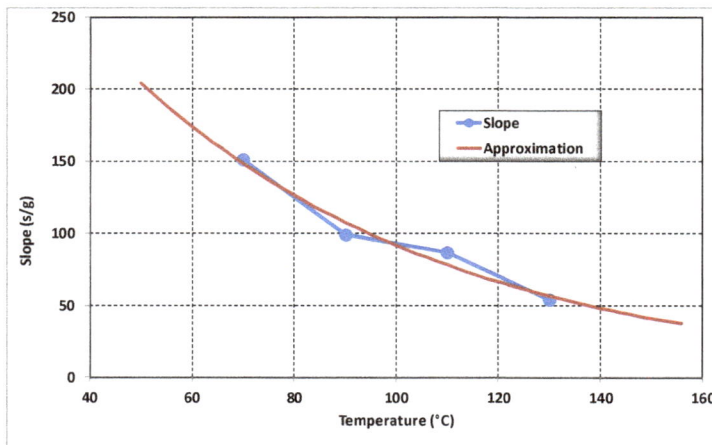

Figure 7. The approximation function of the slopes of Figures 6 and 8 as a function of temperature.

The values of the calculated coefficients K_1 and K_2 are:

$$K_1 = 455.53$$
$$K_2 = 62.31$$

As a result, Equation (9) can be written as:

$$P = 455.53 \cdot e^{-\frac{T}{62.31}} \tag{13}$$

By substituting Equation (13) into (10) we get the following relationship, where drying time constant τ is expressed as a function of the drying temperature T and the mass of paper to be dried M:

$$\tau = 455.53 \cdot M \cdot e^{-\frac{T}{62.31}} \tag{14}$$

From Equations (10) and (14), the following formula is obtained, allowing us to model the drying curves:

$$H_{ext} = H_0 \left(1 - \exp \left(-\frac{t}{455.53 \cdot M \cdot exp\left(-\frac{T}{62.31}\right)} \right) \right) \tag{15}$$

This mathematical model allows us to express extracted moisture (H_{ext}) as depending on paper weight to be dried (M), drying temperature (T), and initial moisture (H_0).

Whether the drying process is performed by the equipment used in this study, or by the transformer manufacturer, the drying process in both cases is stopped when the extraction rate (expressed by quantity of extracted water per paper amount per unit time) reaches a predefined threshold. When this happens, using the mathematical model proposed for the drying process (Equation (10) or (14)) the extraction rate ($Rate_{ext}$) can be expressed by the derivative of the function of the extracted moisture (H_{ext}):

$$Rate_{ext} = \frac{dH_{ext}}{dt} = \frac{H_0}{\tau} e^{-\frac{t}{\tau}} \tag{16}$$

In this case, from the established model, considering that the initial moisture H_0 was previously estimated and that the extraction rate threshold (in our case it is fixed at 0.02% per min) is known, the drying time necessary to achieve this threshold can be calculated using the following equation:

$$t = -\tau \cdot \ln\left(\frac{\tau \cdot rate_{ext}}{H_0}\right) \tag{17}$$

Now, replacing the time constant by its expression (Equation (14)), the following equation is obtained:

$$t = -455.53 \cdot M \cdot \exp\left(-\frac{T}{62.31}\right) \cdot \ln\left(\frac{455.53 \cdot M \cdot \exp\left(-\frac{T}{62.31}\right) \cdot rate_{ext}}{H_0}\right) \tag{18}$$

5.4. Validity of the Established Drying Model

The drying times calculated by the model, based on the paper weight, drying temperature, and drying time determined by the drying apparatus, are shown in Table 3 and Figure 8:

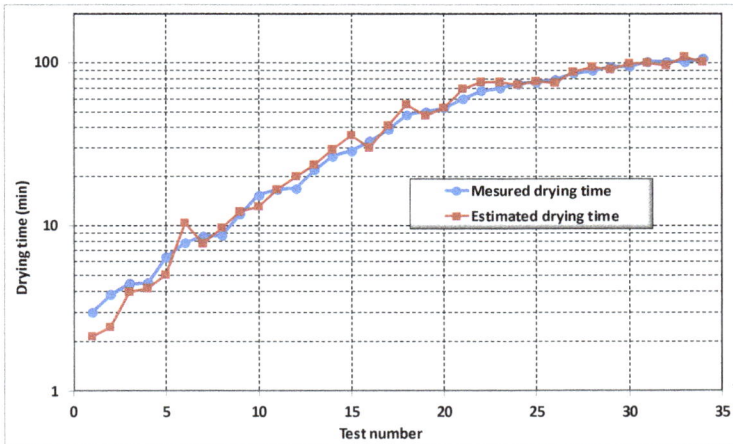

Figure 8. Drying time measured and those determined by the established mathematical model for different drying tests.

Table 3. Drying times measured and those determined by the mathematical model established for different drying tests.

Weight (g)	Measured Duration (min)	Calculated Duration (min)
Drying Temperature: 70 °C		
0.658	7.80	10.32
1.258	15.40	13.04
2.558	32.80	29.80
5.288	50.00	47.24
10.286	78.90	75.06
20.808	94.20	90.42
25.642	101.00	99.16
30.634	101.00	96.24
35.096	105.00	101.23

Table 3. *Cont.*

Weight (g)	Measured Duration (min)	Calculated Duration (min)
Drying Temperature: 90 °C		
0.63	4.50	4.15
1.282	8.60	7.74
2.53	16.70	16.74
4.912	26.60	29.15
9.892	52.70	53.20
20.348	74.50	73.17
31.22	86.40	87.85
35.856	94.80	98.30
Drying Temperature: 110 °C		
0.658	3.82	2.42
1.268	6.40	5.00
2.514	11.73	12.16
5.168	22.00	23.65
10	38.53	40.76
20.11	67.62	75.81
30.11	89.50	93.56
35.432	101.00	108.31
Drying Temperature: 130 °C		
0.638	2.98	2.14
1.29	4.43	3.96
2.454	8.65	9.65
4.978	16.90	20.03
9.942	28.58	35.44
20.094	47.70	55.69
30.042	60.20	68.94
35.45	70.00	76.37
40.13	76.20	77.20

The curves displayed in Figure 8 shows a strong correlation between the calculated and measured durations, which confirms the validity of the proposed drying model. This model can confidently be used for calculating the drying time of the paper used in transformers.

6. Conclusions

Estimating residual moisture is an important factor to ensure acceptable residual water content and avoid failure during dielectric testing certification after manufacture. The success of this estimation depends strongly on the mass of the paper to be dried, the duration of drying, and the drying temperature.

To the best of our knowledge, this estimation is currently made by making an indirect measurement of the extracted moisture. It is thus necessary to use, instead, an equation adapted to the process of drying the paper.

Based on the measurements made on paper samples using the thermogravimetric method, it was possible to set up a model of equations adapted to the paper drying process specific to the transformer. In this model, the drying process of the paper is a function of the temperature, mass of the paper to be dried, initial moisture, and final moisture to be reached, which is the key point of our contribution to the process of drying transformers.

The results obtained in this research show that it is possible to model the phenomenon of drying the paper used for the isolation of power transformers. From this model, it was possible to calculate, for a given threshold of extraction rate and initial moisture, the drying time required as a function of the mass of the paper to be dried, the drying temperature, and the final moisture to be reached.

It was also shown that it would be possible to estimate the residual moisture remaining in the paper at the end of the drying process. The proposed model of paper drying, which is validated against thermogravimetric experimental analyses, has limited applications, because it does not take into account the real conditions of cellulose insulation drying.

This work is the first part of a general model of paper drying for power transformers. Indeed, since the present work was carried out on an equipment with a maximum drying capacity of 50 g, a coefficient must be introduced in the formula in order to adapt the results obtained in this study to the case of large ovens capable of drying large transformers (for example, a 150 MVA transformer may contain as much as 7 tonnes of paper). Moreover, as the drying of transformers is realised under a vacuum, this must also be considered. In the second part of our work, we will take into account these two parameters.

Acknowledgments: This work was carried out within the framework of the Research Chair, ViAHT and the CENGIVRE International Research Center, at Université du Québec à Chicoutimi (UQAC). The authors would like to thank all the sponsors and collaborators.

Author Contributions: This work was done under the supervision of Issouf Fofana, responsible Research Chair on the Aging of Power Network Infrastructure (ViAHT), and Zie Yeo at INP-HB (Institut National Polytechnique Houphouët Boigny Département Génie Électrique et Électronique), who designed this research and gave the whole guidance. Amidou Betie and Fethi Meghnefi collected all the data, carried out calculations, displayed the results and analysis, and wrote the manuscript. The final draft of paper was thoroughly reviewed by Issouf Fofana and Zie Yeo. All authors read and approved the final manuscript.

Conflicts of Interest: The authors declare no conflicts of interest.

References

1. N'cho, J.S.; Fofana, I.; Hadjadj, Y.; Beroual, A. Review of Physicochemical-Based Diagnostic Techniques for Assessing Insulation Condition in Aged Transformers. *Energies* **2016**, *9*, 367. [CrossRef]
2. Fofana, I.; Hadjadj, Y. Electrical-Based Diagnostic Techniques for Assessing Insulation Condition in Aged Transformers. *Energies* **2016**, *9*, 679. [CrossRef]
3. William, H.; Bartley, P.E. Analysis of Transformer Failures. In Proceedings of the International Association of Engineering Insurers 36th Annual Conference, Stockholm, Sweden, 15–17 September 2003.
4. Oommen, T.V.; Prevost, T.A. Cellulose insulation in oil-filled power transformers: Part II maintaining insulation integrity and life. *IEEE Electr. Insul. Mag.* **2006**, *22*, 5–14. [CrossRef]
5. Heathcote, M.J. *A Practical Technology of the Power Transformer*, 13th ed.; Elsevier Ltd.: Oxford, UK, 2007.
6. Su, Q.; James, R.E. *Condition Assessment of High Voltage Insulation in Power System Equipment*; IET Power and Energy Series; Institution of Engineering and Technology (IET): London, UK, 2008; Volume 53.
7. Pahlavanpour, B.; Eklund, M. Thermal ageing of mineral insulating oil and krafts paper. In Proceedings of the TechCon® 2003 Asia-Pacific Conference, Sidney, Australia, 7–9 May 2003.
8. Fournié, R. *Les Isolants en Électrotechnique: Essais, Mécanismes de Dégradation, Applications Industrielles*; Eyrolles: Paris, France, 1990.
9. Garcia, D.F.; Garcia, B.; Burgos, J.C. A review of moisture diffusion coefficients in transformer solid insulation-part 1: Coefficients for paper and pressboard. *IEEE Electr. Insul. Mag.* **2013**, *29*, 46–54. [CrossRef]
10. Sparling, B.; Aubin, J. Assessing Water Content in Insulating Paper of Power Transformers. Available online: http://www.electricenergyonline.com/show_article.php?mag=44&article=333 (accessed on 21 July 2015).
11. Nikjoo, R. *Diagnostics of Oil-Impregnated Paper Insulation Systems by Utilizing Lightning and Switching Transients*; KTH School of Electrical Engineering: Stockholm, Sweden, 2014.
12. Zabeschek, S.; Strzala, H. Drying of High Voltage Power Transformers in the Field with a Mobile Vapour Phase Drying Equipment. Available online: http://www.weidmann-solutions.cn/huiyi/Seminar%202007%20Florida/2007zabeschek.pdf (accessed on 26 February 2018).
13. Tariq, M. Estimating moisture in Power Transformers. In Proceedings of the Transformer Life Management Conference, Dubai, UAE, 22–23 October 2013.
14. Lundgaard, L.E.; Hansen, W.; Linhjell, D.; Painter, T.J. Aging of oil-impregnated paper in power transformers. *IEEE Trans. Power Deliv.* **2004**, *19*, 230–239. [CrossRef]

15. Bartels, M. We have the answer—Precise monitoring of the drying process for power transformers. *Omicron Mag.* **2012**, *3*, 40.
16. Gavrilovs, G.; Sandra, V. Solid insulation drying of 110 kV paper-oil instrument transformers. *Sci. J. Riga Tech. Univ. Power Electr. Eng.* **2009**, *25*, 35–38. [CrossRef]
17. Du, Y.; Zahn, M.; Lesieutre, B.C.; Mamishev, A.V.; Lindgren, S.R. Moisture equilibrium in transformer paper-oil systems. *IEEE Electr. Insul. Mag.* **1999**, *15*, 11–20. [CrossRef]
18. Hall, C.W. The evolution and utilization of mathematical models for drying. *Math. Model.* **1987**, *8*, 1–6. [CrossRef]
19. Doymaz, I.; Pala, M. The effects of dipping pretreatments on air-drying rates of the seedless grapes. *J. Food Eng.* **2002**, *52*, 413–417. [CrossRef]
20. Chayjan, R.A.; Salari, K.; Abedi, Q.; Sabziparvar, A.A. Modeling moisture diffusivity, activation energy and specific energy consumption of squash seeds in a semi fluidized and fluidized bed drying. *J. Food Sci. Technol.* **2013**, *50*, 667–677. [CrossRef] [PubMed]
21. García, D.F.; García, B.; Burgos, J.C.; García-Hernando, N. Determination of moisture diffusion coefficient in transformer paper using thermogravimetric analysis. *Int. J. Heat Mass Transf.* **2012**, *55*, 1066–1075. [CrossRef]
22. Garcia, D.F.; Garcia, B.; Burgos, J.C.; Hernando, N.G. Experimental determination of the diffusion coefficient of water in transformer solid insulation. *IEEE Trans. Dielectr. Electr. Insul.* **2012**, *19*, 427–433. [CrossRef]
23. Midilli, A.; Kucuk, H.; Yapar, Z. A New Model forsingle-layer drying. *Dry. Technol.* **2002**, *20*, 1503–1513. [CrossRef]
24. A&D Company Limited. *A&D's Moisture Analyzers Instruction Manual*; A&D Company Limited: Tokyo, Japan, 2004.

Article

Dual-Temperature Evaluation of a High-Temperature Insulation System for Liquid-Immersed Transformer

Xiaojing Zhang [1], Lu Ren [1], Haichuan Yu [1], Yang Xu [1,*], Qingquan Lei [1], Xin Li [2] and Baojia Han [3]

[1] State Key Laboratory of Electrical Insulation and Power Equipment, Xi'an Jiaotong University, No. 28 Xianning West Road, Xi'an 710049, China; Emma.Zhang@dupont.com (X.Z.); renluxjtu@163.com (L.R.); hnbcyhc@stu.xjtu.edu.cn (H.Y.); lei_qingquan@sina.com (Q.L.)

[2] Electric Power Research Institute of Guangdong Power Grid Corporation, No.6–8 Dongfeng East Road, Guangzhou 510062, China; lihongxin0303@163.com

[3] Guangdong Zhongpeng Electricity Co., Ltd., No.19 Qianjin Road, Foshan 528000, China; hanbaojia@hotmail.com

* Correspondence: xuyang@xjtu.edu.cn; Tel.: +86-029-82665415

Received: 5 July 2018; Accepted: 24 July 2018; Published: 27 July 2018

Abstract: A high-temperature oil–paper insulation system offers an opportunity to improve the overloading capability of distribution transformers facing seasonal load variation. A high-temperature electrical insulation system (EIS) was chosen due to thermal calculation based on a typical loading curve on the China Southern Power Grid. In order to evaluate candidate high-temperature insulation systems, Nomex® T910 (aramid-enhanced cellulose) immersed in FR3 (natural ester) was investigated by a dual-temperature thermal aging test compared with a conventional insulation system, Kraft paper impregnated with mineral oil. Throughout the thermal aging test, mechanical, chemical, and dielectric parameters of both paper and insulating oil were investigated in each aging cycle. The thermal aging results determined that the thermal class of the FR3-T910 insulation system meets the request of overloading transformer needs.

Keywords: liquid-immersed transformer; thermal aging; Nomex T910; natural ester; dual-temperature

1. Introduction

Urbanization in China has led to the population moving between urban and rural areas, which results in seasonally fluctuating electrical power demand, particularly overloading rural transformers during the Chinese spring festival and plowing seasons. Traditionally, the adoption of higher-capacity transformers was considered by utilities to handle overloading. However, this decreases the normal loading rate and yields a lower efficiency.

The State Grid Corporation of China published technical guidance for overloading rural transformers in 2014 [1], and identified the typical loading curve. China Southern Power Grid released a similar typical loading curve, shown in Figure 1 [2]. On the basis of the normal temperature-rise test requirements, the transformer should meet the requirement for continuous operation of 6 h at 1.5 times rated power, 3 h at 1.75 times rated power, and 1 h at 2 times rated power. A conventional insulation system, Kraft paper and mineral oil, has been utilized as a principal insulation system in liquid-immersed transformers. A typical thermal class of Kraft paper is 105 °C, which allows an average winding temperature rise of 55 K [3]. Meanwhile, the flash point of mineral oil is about 140 °C, which raises fire safety concerns for the industry.

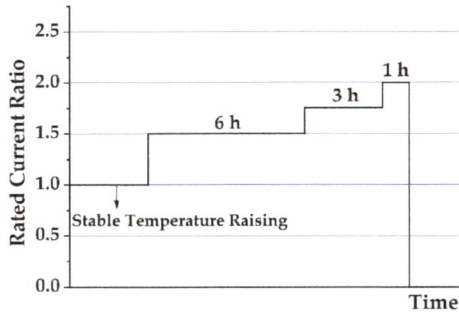

Figure 1. Loading curve for overloading distribution transformers required by Southern Grid [2].

The sustainable solution is to improve the overloading capacity of distribution transformers by adopting a thermally upgraded insulation system. Natural and synthesized esters have been developed in the past, focusing on fire safety and environmental footprint. Several researchers [4–6] have reported that natural ester is beneficial in alleviating the deterioration of paper and prolonging its life, due to the high hydrophilicity and the transesterification effect on cellulose of natural ester. Based on the thermal calculation referring to a typical loading curve, the maximum hot spot temperature is 117.9 °C for the distribution transformer (S13-m(b)-400/10). In order to meet the high-temperature needs of insulation paper, Nomex® T910 was introduced globally in 2014 as an alternative insulating paper for liquid-immersed transformers.

As shown in Figure 2, Nomex® T910 is a type of aramid-enhanced cellulose paper with a three-layer sandwich structure. The center ply is made of 100% cellulose, like normal Kraft paper, but the outer two plies are made of 50–70% cellulose blended with 30–50% aramid materials. Three plies are consolidated into a single sheet by a paper manufacturing processes such as wet forming, drying, and densification. The reason why T910 has higher thermal performance than cellulose paper is that the aramid fiber can enhance the thermal stability of the cellulose paper, and the fiber ratio between aramid fiber and cellulose fiber will determine its performance. The dispersal states will be different when the ratios of aramid fiber and cellulose fiber are different, as shown in Figure 2.

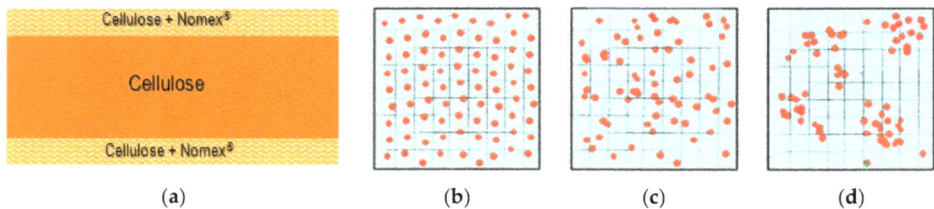

Figure 2. The structure of T910 [7] and different dispersal states of aramid fiber in cellulose fiber under different ratios: (**a**) structure of T910; (**b**) regular, with ratio less than 5%; (**c**) random, with ratio between 10% and 25%; (**d**) clumped, with ratio higher than 30%.

Once the ratio of aramid fiber is lower than 5%, the thermal or electrical chambers between modules are similar to those of cellulose paper. If the aramid fiber ratio is increased to 10–25%, cellulose fiber will distribute randomly in the paper, which improves the paper's performance. Furthermore, when the aramid fiber ratio is increased to 30%, cellulose fibers will be surrounded by aramid fiber. The traditional thermal chambers between cellulose fibers have been blocked to improve the thermal performance of paper significantly, and the final content of aramid fiber in the center ply is about

30–50%. As shown in Figure 3, the Thermal Gravimetric Analyzer (TGA) results of both Kraft and T910 have been analyzed. Although the initial decomposition temperature of T910 is 310.87 °C, the secondary decomposition of T910 has been improved to 529.32 °C, compared with Kraft by aramid fibers, which potentially bring the benefit of high temperature thermal stability.

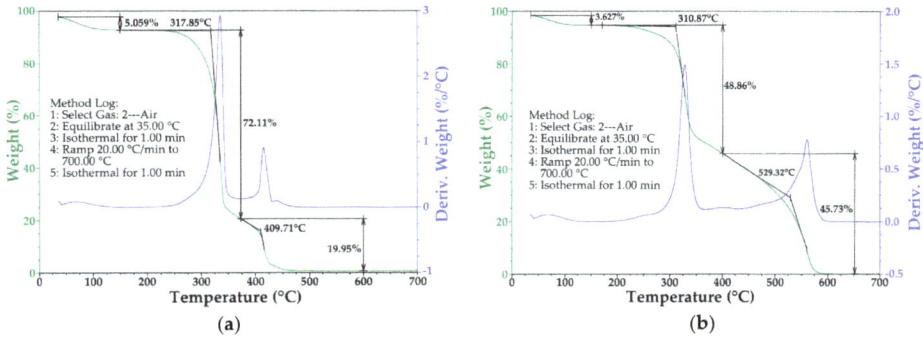

Figure 3. TGA results of insulating papers: (**a**) cellulose paper; (**b**) T910.

Before applying a new insulation system to the transformer, a thermal aging test should be done to test its thermal class or thermal index. According to [7], the thermal class of mineral oil–impregnated T910 is considered to be 130 °C. The tensile strength of mineral oil decreases slightly after 250 h thermal aging under 150 °C, or 500 h thermal aging under 130 °C [8]. However, there are no details or results on the thermal aging of natural ester impregnated T910.

Three accelerated thermal aging methodologies can be adopted to evaluate the thermal aging performance of a new insulation system, as follows. IEEE Std. C57.100^{TM} [9] gives a description of functional life aging, called the Lockie method, which is closer to the real operation by introducing a distribution or power transformer model with high cost. Another approach, the sealed tube method, is commonly adopted due to its convenience; this method has been described in IEC 62332-2 [10]. The latest aging method, the dual-temperature test, was introduced in IEC 62332-1 [11], and simulates operational status, including the different temperatures between insulating paper and oil as well as oil flow, which are the limitations of seal tube tests. According to research by Wicks [12], the two evaluations, the dual-temperature aging method and the Lockie method, are similar for the same insulation system.

In this paper, in order to evaluate the thermal aging performance of T910-FR3 and Kraft paper-mineral oil, thermal aging tests were conducted by using the dual-temperature platform, which was adjusted to simulate real transformer winding compared to the structure of IEC 62332-1.

2. Experimental Methods

2.1. Materials

The insulating paper tested for this study included Nomex® T910 paper (0.08 mm) manufactured by DuPont and Kraft paper (0.08 mm) by Nine Dragons Paper, as well as high-density pressboard (1.17 g/cm^3). Insulating oil included FR3 from Cargill and No. 25 mineral oil (MO) from Nan You Petrochemical. The basic properties of the different oils and papers are shown in Table 1.

Table 1. Basic properties of insulating oils and insulating papers.

Properties	Insulating Oil		Properties	Insulating Paper	
	MO	FR3		Kraft	T910
Appearance	Clear	Light Green	Thickness (mm)	0.08	0.08
Density, 20 °C, (g·cm^{-3})	0.848	0.92	Density, (g·cm^{-3})	0.89	0.9–1.1
Viscosity, 40 °C, (mm^{-2}·s^{-1})	8.01	32–34	Tensile Strength, MD (N·cm^{-1})	64.7	70
Flash Point, PMCC (°C)	155	255	Tensile Strength, XD (N·cm^{-1})	38.6	17
Acid Value (mg KOH·g^{-1})	0.0011	0.013–0.042	AC Breakdown Voltage, Air (kV·mm^{-1})	10.4	20
Dissipation Factor (90 °C)	0.00101	0.02	Dissipation Factor (%)	0.25 (50 Hz, 100 °C)	1.6 (60 Hz, 90 °C)

2.2. Dual-Temperature Thermal Aging Platform

As shown in Figure 4, the dual-temperature test platform includes an aging cell, power supply system, heating elements, control system, safety protector, and sampling unit. The aging cells in the tests were modified based on IEC 62332-1 [11], as labeled in Figure 5. The outer diameter of the cell is 220 mm and the length is 420 mm. The aging cell consists of both an insulating oil–paper test object sample and corresponding heaters and sensors, in which temperature monitoring of solid and liquid materials is separately realized by the copper conductor and immersion heaters coordinating with their individual thermal sensors. The insulating oil circulates in the aging cell due to the temperature difference to simulate the oil flow in transformers.

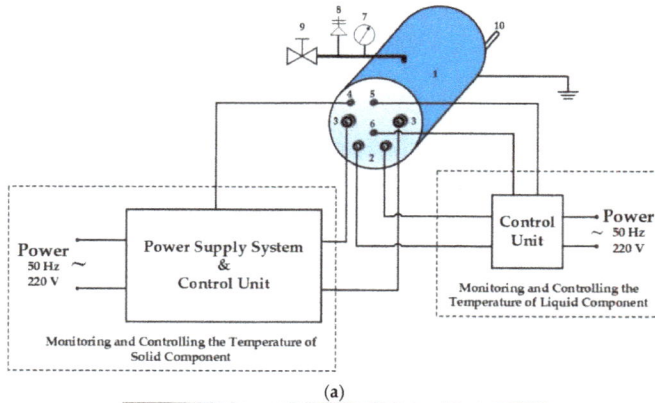

(a)

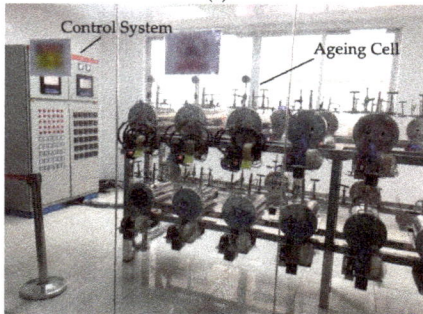

(b)

Figure 4. Dual-temperature thermal aging platform: (**a**) schematic diagram (1, aging cell; 2, immersion heaters for liquid; 3, copper conductor; 4, temperature sensor for solid component; 5, temperature sensor located at liquid surface; 6, temperature sensor located at liquid center; 7, barometer; 8, relief valve; 9, gas valve; 10, liquid pipeline); (**b**) photo.

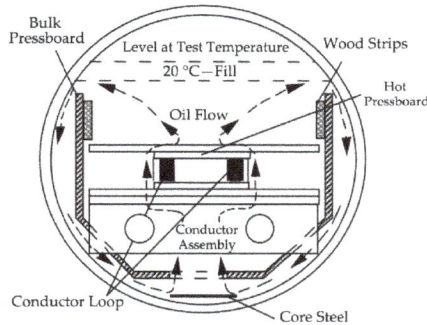

Figure 5. Sectional view of aging cell [11].

The solid samples are mainly composed of a copper conductor, the insulating papers to be evaluated, and a pressboard. The copper conductor, made into an M type, is used to simulate the winding of the transformer, which is first wrapped with paper tape in a half-lapping way to prevent contamination of test samples, then by tensile and breakdown samples in sequence, and finally, by paper tape, as in the first step for fixing. The M type copper conductor wrapped with paper is tightly bound with copper wires and high-density pressboard, in order to simulate transformer winding conditions. Considering the temperature differences among the 4 parts of the conductor, test samples are classified into outer layer samples and inner layer samples. There are 7 layers for both inner layer and outer layer: 1 layer for foundation, 5 layers for tensile strength, breakdown voltage, and initial tear strength test, and 1 layer for protection. The difference between them is the temperature difference based on thermal radiation. Thermal sensors are located between the first layer and tensile samples, in order to reflect the temperature of the tensile samples. As shown in the internal view of an aging cell in Figure 6, the copper conductor is fixed to the lid of the aging cell, and connected to the power system using fluororubber for sealing and insulation.

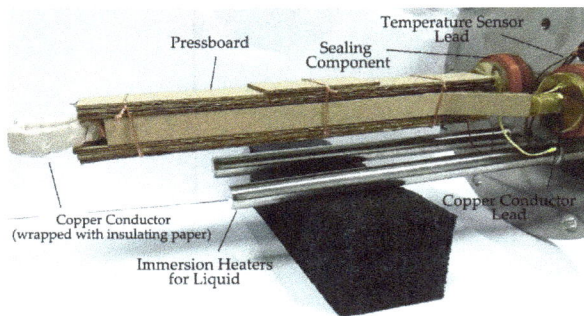

Figure 6. Sample of internal view of aging cell.

To make the test cell reasonable, the volume ratios of materials are selected according to the recommendations of IEC 62332-1 [11]. In this test, the oil sample consumption is about 10.3 L. The calculation allowed the determination of material volume ratios as follows: 474 cm^3 of high-temperature insulation (which includes the insulation and the pressboard in high-temperature areas), 1140 cm^3 of low-temperature insulation (which includes low-temperature pressboard and additional samples for material balance), and 214 cm^2 of silicon steel. Specifically, 474 cm^3 high-temperature insulation includes 18 cm^3 candidate insulating paper sample, 115 cm^3 of paper for the wrapped conductor, and 341 cm^3 of high-density pressboard. The thickness of all paper for wrapped conductors is 0.08 mm. In this paper, 2 insulation systems were tested: Kraft paper–mineral oil as the benchmark system, and T910-FR3 ester as the candidate system. Both of them had the same volume ratio of insulating paper and oil, but different mass ratio based on different densities. Each aging cell includes 118 g of Kraft or 133 g of T910, 399 g of high-density pressboard for high temperature insulation, and 510 g of high-density pressboard for the low temperature pressboard. For insulating oil, mineral oil was 8.73 kg in the conventional insulation system, and FR3 was 9.48 kg in the high-temperature insulation system.

2.3. Sample Pretreatment

All the insulating oil and papers should be preconditioned before the test. The insulating oil should be dehydrated and degassed under vacuum conditions by using the apparatus shown in [13]. At the same time, the insulating paper should be dehydrated under high temperature. Figure 6 shows the assembly of the insulation paper, high-density pressboard, and other accessories. As the last step in the dehydration process, the aging cell was put into the oven with open valves after sealing. As the preconditioning result, water contents of the materials are listed in Table 2.

Table 2. Water contents of insulating oil and paper after preconditioning.

Property	Insulating Oil		Insulating Paper	
	MO	FR3	Kraft	T910
Water Content	17.1 ppm	147.4 ppm	0.503%	0.426%

Furthermore, the pipeline of the aging cell was connected to the processed oil, and the gas valve was connected to the vacuum pump to keep the low pressure of the aging cell. Insulating oil in the amount of 10.3 L was injected into the aging cell. After the oil injection and vacuum process, the valves were closed and the aging cell was put into the oven. The solid samples were impregnated with corresponding insulating oil at 90 °C for 12 h under vacuum conditions to soak in fully. After the impregnation process, the absorption of T910 with FR3 was 42.1%, and that of Kraft paper with mineral oil was 37.5%, based on the density difference and fiber polarity variance of the insulating paper.

2.4. Thermal Aging Test Conditions: Temperature and Aging Period

To confirm the thermal performance of the candidate system, life assessments of both the candidate and referenced systems at 3 temperatures are necessary, according to the Arrhenius equation. In this paper, the aging temperatures of the copper conductor and insulating oil were separately controlled to simulate hot spot temperature and top oil temperature differences. Details about the aging temperatures and periods of the candidate system and referenced system refer to IEC 62332-1 [11] and are listed in Table 3. After each aging period, the tensile strength and breakdown voltage of insulating paper are measured, as well as water content, acid number, viscosity, and breakdown voltage of insulating oil.

Table 3. Insulation system thermal aging conditions.

Reference EIS (MO and Kraft)				Candidate EIS (FR3 and T910)			
Temp. of Kraft (°C)	Temp. of Mineral Oil (°C)		Aging Cycle (h)	Temp. of T910 (°C)	Temp. of Natural Ester (°C)		Aging Cycle (h)
	Surface	Center			Surface	Center	
160	124	115	25/50/100/250	180	130	115	100/250/500/1000
140	122	115	250/500/2000/4000/5000	165	127	115	500/2000/4000/5000
125	121	115	3000/4500/5000	150	125	115	3000/4500/5000

3. Aging Test Results of Insulating Papers

3.1. Tensile Strength of Insulating Papers

Cellulose is a kind of linear condensation polymer consisting of anhydroglucose; the degree of polymerization (DP) is the average number of glycosidic rings in a cellulose macromolecule [3]. The tensile strength of insulating paper has a direct correlation with DP. When the DP of cellulose paper decreases, the tensile strength will also decrease [14]. In addition, the insulating paper has to endure a certain degree of mechanical stress during the operation of the transformer. Once mechanical fractures appear in the insulating paper, the winding will be directly exposed to the insulating oil, and electrical faults are more likely to occur. Moreover, the locations of fractures can also cause partial discharge as insulating defects. So, the tensile strength of insulating paper is commonly used as an index to evaluate the aging status of the paper [15]. In this study, the tensile strength of insulating paper with different aging statuses was measured according to ASTM D828 [16].

After preconditioning, the initial tensile strength of Kraft paper impregnated with mineral oil was 79.0 N/cm, and of T910 impregnated with FR3 ester was 84.7 N/cm. Residual tensile strength was selected to characterize aging status in insulating paper; the calculating method is shown as Equation (1):

$$RTS = \frac{TS}{ITS} \times 100\% \tag{1}$$

where *RTS* is residual tensile strength, *TS* is tensile strength after different aging periods, and *ITS* is the initial tensile strength of the corresponding paper. The results reflect the trend of *RTS* with aging time, according to IEC 60216-3 [17]. Fifty percent of *ITS* mentioned in IEC 60216-2 [18] is benchmarked as the lifetime guideline of insulating papers. Figure 7 shows the results of *RTS* of Kraft and T910 paper at different aging temperatures for the inner and outer layers.

The initial tensile strength of each sample has been treated as 100% as the initial point before thermal aging process. During the first stage of each aging test, the *RTS* between the initial point and the first measured point dropped apparently, even more with an increase in temperature. For the comparison between the two layers, the *RTS* of the inner layer dropped more than that of the outer layer. Furthermore, the scatter points without the initial one are fitted linearly by the method of least squares to determine the lifetime of each insulation system under different aging conditions. As normal practice, the lifetime will be considered as the ageing time when the tensile strength dropped to 50% of the initial tensile strength. The results of lifetime under each aging temperature are listed in Table 4. The lifetime of the outer insulating paper is higher than that of the inner layer under the same aging temperature due to the difference of heat dissipation situation. The outer layer contacted with insulating oil directly, which leads to a better heat dissipation situation. As expected, the lifetime decreases with the increase of the aging temperature for both insulation systems. However, the lifetime of T910 is still much higher than that of Kraft paper, even under higher aging temperature, which illustrates a relatively better thermal capability.

Table 4. Life (in hours) of insulating paper at each aging temperature.

Life of Kraft (h)	160 °C	140 °C	125 °C
Outer Layer	138	1734	4044
Inner Layer	38	804	2718
Life of T910 (h)	**180 °C**	**165 °C**	**150 °C**
Outer Layer	576	2735	4878
Inner Layer	259	1583	3408

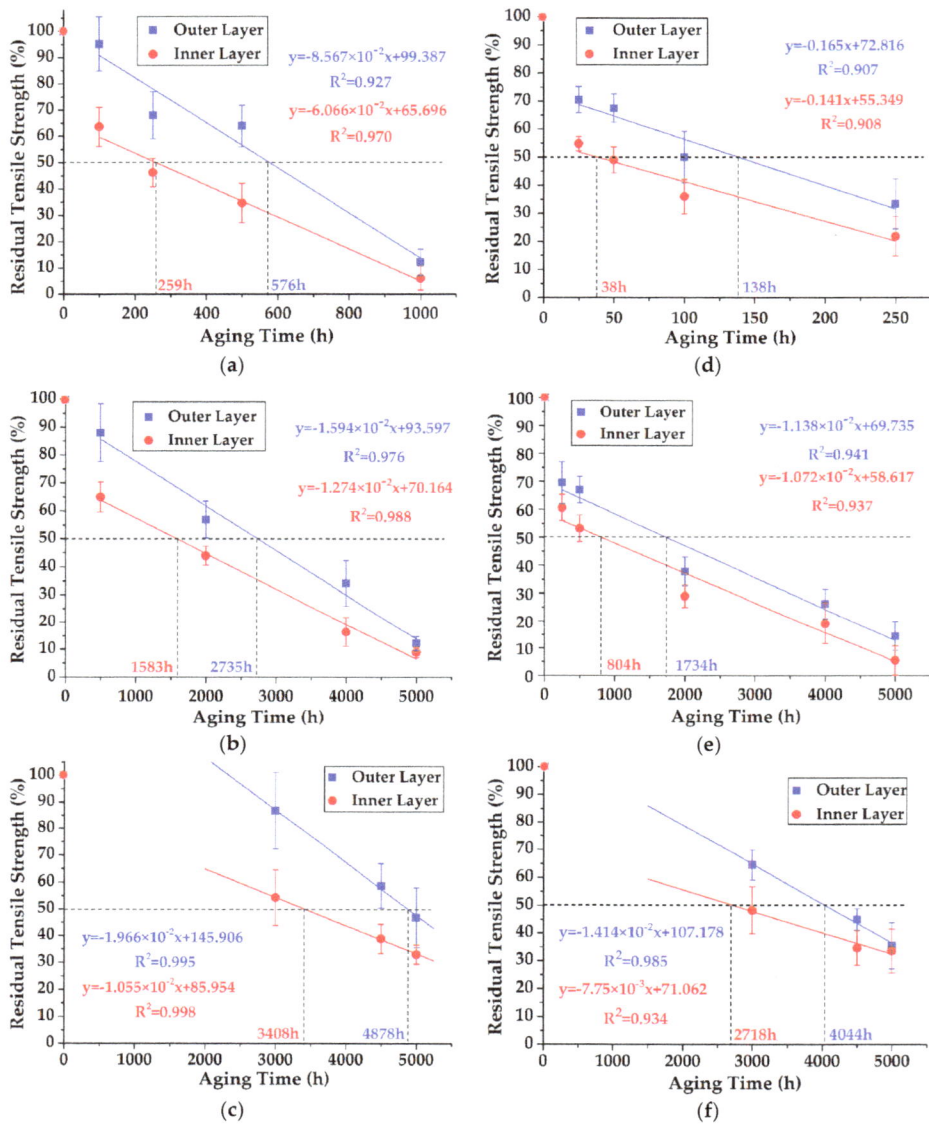

Figure 7. Residual tensile strength of T910 in FR3 and Kraft in mineral oil at different temperatures: (**a**) T910 at 180 °C; (**b**) T910 at 165 °C; (**c**) T910 at 150 °C; (**d**) Kraft at 160 °C; (**e**) Kraft at 140 °C; (**f**) Kraft at 125 °C.

3.2. Thermal Index (TI) of Insulation Systems

According to the Arrhenius equation, there is a relationship between the chemical reaction rate and thermodynamic temperature. Thus, the relationship between the thermal aging lifetime and aging temperature of an insulation system can be given as Equation (2):

$$\text{Log } L = A + \frac{B}{T + 273} \tag{2}$$

where L represents the lifetime of a certain insulation system in hours, T represents temperature in °C, and A and B are constants. According to Equation (2), the thermal aging life curves of two insulation systems are plotted in Figure 8.

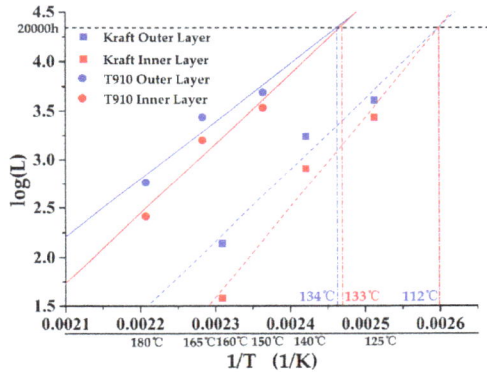

Figure 8. Thermal aging life curves of insulation systems.

Comparing the two different positions, the heat dissipation condition of the outer layer paper is better than that of the inner layer paper, which leads to the lifespan of the outer layer paper being relatively higher under the same temperature. However, the gap of thermal aging life curves between inner and outer layer samples was reduced under low thermal aging temperatures. Normally, the intermediate oil temperature is controlled at 115 °C, which is lower than the aging temperature of paper. Once the thermal aging temperature of insulating paper is low, the temperature difference between insulating paper and oil decreases. Furthermore, the thermal environment of the insulation system inside and outside tends to be consistent. The slope of the life curve of the outer layer paper is less than the life curve of the inner layer paper, and finally, intersects along with the decrease of thermal aging temperature.

TI refers to the value of Celsius temperature when the lifetime curve is 20,000 h. The TI of inner layer T910 impregnated with FR3 is 133 °C, lower than that of the outer layer at 134 °C. In the MO impregnated Kraft insulation system, the TI of the inner and outer layer paper is approximately 112 °C. The heat resistance of the FR3 impregnated T910 insulation system is much higher than that of the MO impregnated Kraft insulation system.

3.3. Dielectric Strength of Insulating Paper

Dielectric strength is also one of the important properties of insulation paper, which directly provides insulation protection for winding. The breakdown voltages of insulating paper under different temperatures in different aging cycles were measured and plotted, as shown in Figure 9. Since the insulation paper was wrapped on a copper conductor, creases are unavoidable on its surface. In order to reduce the influence of creases on the breakdown voltage of insulation paper, the VDE electrode was adopted for step-up voltage tests with 50 Hz power supply frequency and 0.5 kV/s step-up speed. In these tests, T910 was placed in preconditioned FR3, and Kraft was placed in preconditioned mineral oil. The breakdown voltage of each sample was measured 20 times. All results were plotted by Weibull distribution, and the value with 63.2% breakdown probability was considered as the final result. Considering that the thickness of both candidate insulating papers are the same, their breakdown voltages can be directly compared.

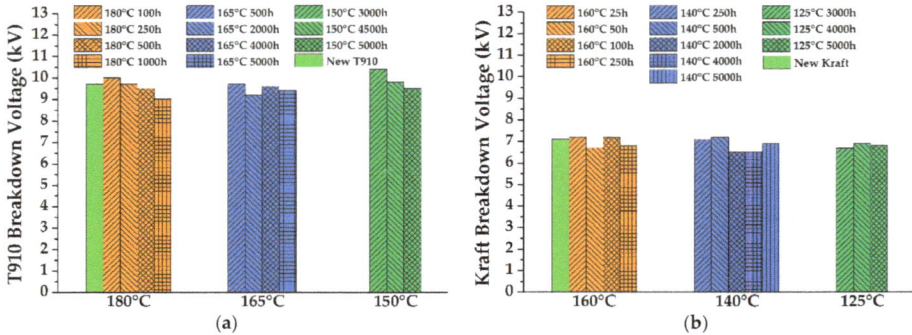

Figure 9. Breakdown voltage of aged insulating paper: (**a**) aged T910; (**b**) aged Kraft paper.

For the initial value of new insulating paper, T910 has better dielectric property breakdown voltage (T910: 9.7 kV; Kraft paper: 7.1 kV). Compared with the new paper, the maximum voltage drops of T910 and Kraft paper after aging is 0.7 kV and 0.6 kV respectively, which is not significant for either. Different from tensile strength, breakdown voltage is stable throughout the aging process. Reference [19] also reported that even if the mechanical strength of insulating paper drops to 30% of its initial value, dielectric strength can still maintain a relatively high level.

4. Aging Test Results of Insulating Oils

4.1. Water Content

Water content is one of the most important factors in the dielectrical performance of insulating oil [20], while water modules will accelerate the aging process for oil–paper. Coulometric Karl Fisher titration methodology was adopted in this study by Metrohm 831 KF coulometer and 860 KF Thermoprep. Considering similar trends under different temperatures, a typical thermal aging temperature was chosen to be measured for each insulation system, as shown in Figure 10 (aged oil under temperatures of 165 °C and 140 °C for T910 and Kraft, respectively).

For mineral oil, the water content increased continuously throughout the aging process. The initial water content of FR3 was higher than that of mineral oil. However, the water content of FR3 showed different trends, increasing to the peak during the beginning of the aging process but dropping later. The water content in FR3 at the end of the aging test was even lower than at the start, which means that the water was consumed during the process. Similar changing of water content is shown in [21,22], which suggests an accelerated aging rate in mineral oil compared to FR3, because hydrolysis, for which water is the reactant, is the main cause cellulose degradation of both Kraft and T910 [23]. Aramid has excellent chemical stability and is resistant to hydrolysis and oxidation, whose by-products, gas and water, are lower than the cellulose [24]. Paper [25] has illustrated that even under 240 °C, aramid paper still retained 77% tensile strength in mineral oil after 5000 h thermal aging test. In addition, the maximum aging temperature within the tests is 180 °C, which is lower than pyrolysis temperature of aramid. As Figure 2 shows, cellulose fiber has been surrounded by aramid fiber to enhance the thermal resistance performance.

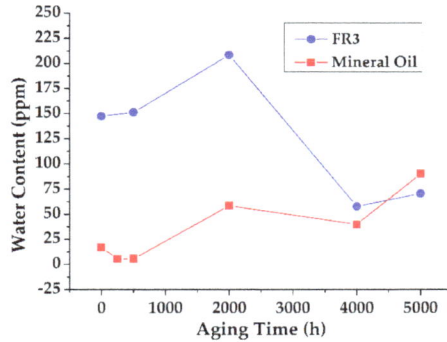

Figure 10. Water content of insulating oil under medium aging temperature (165 °C for T910 aged in FR3 and 140 °C for Kraft aged in mineral oil).

Moreover, compared with mineral oil, FR3 has a higher saturated water content, which means that FR3 can absorb more water and decrease the water content of insulating paper.

4.2. Dynamic Viscosity

Insulating oil must be mobile to transfer the heat in a better manner from the core to the transformer radiators, where heat is dissipated. The dynamic viscosity of insulating oil is determined by relative molecular weight, in a relation shown by Mark-Houwink in the equation:

$$\eta = KM^a \tag{3}$$

where η is dynamic viscosity, a and K are constants depending on the particular polymer–solvent system, and M is the average molecular weight. In this paper, the dynamic viscosity of insulating oil was measured by a Brookfield LV-II pro viscometer fitted with a small sample adapter under 40 °C and 90 °C. Figure 11 shows the viscosity of oil under different thermal aging conditions: 165 °C for T910 and 140 °C for Kraft. The dynamic viscosity of FR3 is significantly higher than that of mineral oil at the two temperatures, based on the relative molecular weight of triglyceride, which is a potential bottleneck for natural ester to be adopted in a liquid-immersed transformer. In this paper, the nitrogen protection of aging cells narrowed the variance of viscosity due to the limited oxygen reaction.

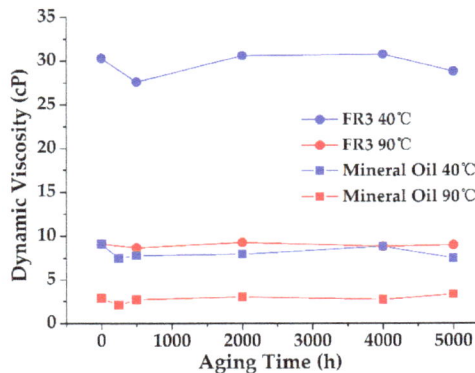

Figure 11. Dynamic viscosity of insulating oil under medium thermal aging temperature (165 °C for T910 and 140 °C for Kraft).

4.3. Total Acidity

The acidity of insulating oil speeds up the aging process of insulating paper, because hydrolysis of cellulose is catalyzed by H⁺ [26]. In this paper, the acidity of insulating oil was measured by potentiometric titration with KOH isopropanol solution by Metrohm 848 Titrino plus equipment (Metrohm, Herisau, Switzerland). Considering the same pattern will be followed for acidity under different thermal aging temperatures, Figure 12 shows the total acidity of insulating oil under a typical aging temperature as an example.

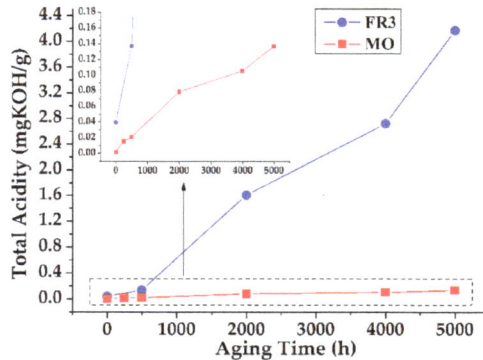

Figure 12. Total acidity of insulating oil under medium thermal aging temperature (165 °C for T910 and 140 °C for Kraft).

All thermal aging tests were conducted under nitrogen protection, with the acidity of insulating oil affected by H⁺ generated from the hydrolysis of Kraft paper. Figure 12 shows the increasing acidity of mineral oil throughout the aging process. Compared with mineral oil, both initial value and speed of increase of the acidity of FR3 are higher due to the generation of fatty acid from the hydrolysis of glycerol fatty acid ester, which is the main component of natural ester. Different from carboxylic acid, fatty acid is a high-molecular acid that will not catalyze the hydrolysis of insulating paper [27]. In addition, a transesterification reaction between fatty acid and cellulose in T910 can form a barrier to water ingress and postpone the degradation of solid insulation [21,22]. In summary, a higher total acidity of FR3 cannot directly reflect the thermal aging status compared with mineral oil.

4.4. Breakdown Voltage

Breakdown voltage tests were conducted with a BAUR DTA 100C breakdown voltage machine (BAUR GmbH, Sulz, Austria), with reference to ASTM D1816 [28]. Step-up voltage tests were conducted with a 50 Hz power supply frequency and 0.5 kV/s step-up speed; the distance between electrodes was 1 mm. For each insulating oil sample, the breakdown voltage tests were conducted 50 times at room temperature after each thermal aging cycle. All results were plotted by Weibull distribution, and breakdown voltage under probability equal to 63.2% was set as a benchmark, as shown in Figure 13.

As Figure 13a shows, the breakdown voltages of insulating oil have dispersion in a range. Based on statistical analysis principles, the value of breakdown voltage is adopted once the probability is 63.2%. As Figure 13b,c shows, the initial breakdown voltage of FR3 is 41.1 kV, and that of mineral oil is 39.4 kV. For FR3, the difference of temperature and aging time has a limited impact except on the breakdown voltage under 180 °C for 1000 h, which is 31.2 kV. However, the lowest value of breakdown voltage is still higher than the controlling value of the related standard. In contrast, for mineral oil, the variance of breakdown voltage under different aging conditions is significant.

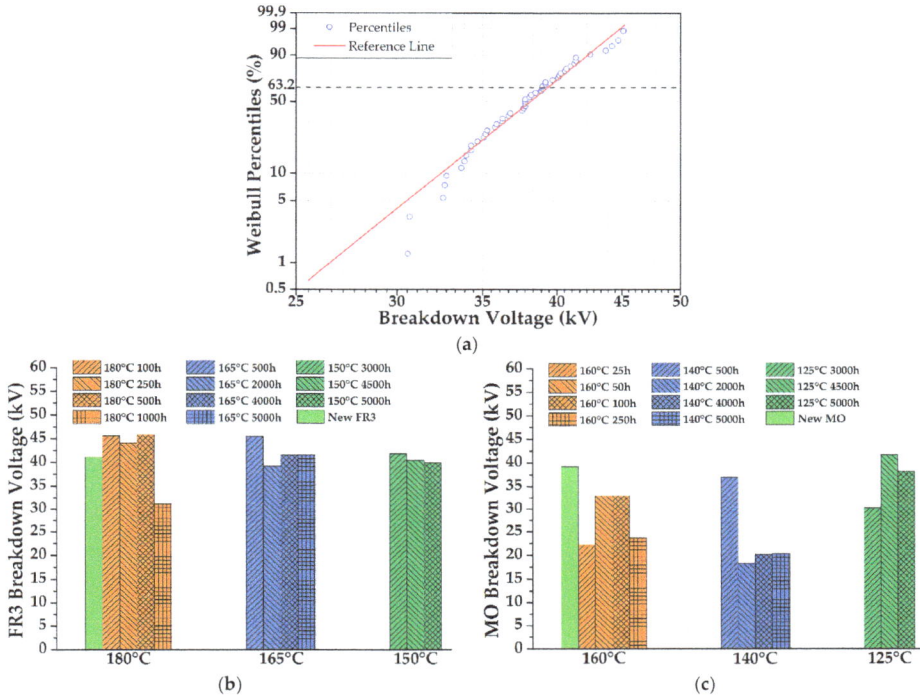

Figure 13. Breakdown voltage of insulating oil under different thermal aging conditions: (**a**) Weibull percentiles of new MO; (**b**) aged FR3; (**c**) aged MO.

According to a related study, the breakdown voltage of mineral oil is more sensitive to water content compared with natural ester [24]. For example, the saturated water content of mineral oil is about 68 ppm, and that of FR3 is about 1100 ppm at 25 °C [29]. The water molecule within the mineral oil will be released once the water content increases to the limit, which brings the potential risk of breakdown. Compared with absolute water content, relative water content is meaningful as a reference that can be calculated by the ratio of water content to saturated water content. With the increase of water content, breakdown voltage dropped significantly for both FR3 and mineral oil [24,30]. From Figure 10, after aging at 165 °C, it was observed that the water content of FR3 increased from 151.2 ppm after 200 h to 208.6 ppm after 2000 h, then decreased to 57.9 ppm after 4000 h. At the same time, the breakdown voltage of FR3 first decreased from 45.5 kV to 39.2 kV and then increased to 41.6 kV.

5. Conclusions

In this paper, a dual-temperature thermal aging platform was established to explore a high-temperature insulation system, Nomex T910 immersed by FR3, as long as the conventional insulation system is identified as the benchmark.

For solid insulation materials, even under higher thermal aging temperatures, the tensile strength of T910 dropped more slowly compared with Kraft paper; that means the thermal index of T910 is higher than that of Kraft paper. In addition, the variance of dielectric strength for solid insulation is smaller, and in particular, T910 showed better performance during the whole thermal aging process.

For fluids, the viscosity of both mineral oil and FR3 maintains a relatively similar level, which shows that the oxygen reaction between solid insulation and fluids is not evident under nitrogen protection. The acidity of FR3 is higher compared with mineral oil based on chemical composition. Fatty acid was generated by glycerol fatty acid ester after hydrolysis during the thermal aging process. However, esterification between fatty acid and cellulose will postpone the degradation of cellulose opposite to other acids. After thermal aging, the breakdown voltage of FR3 maintained a high level, except after 1000 h at 180 °C, but was still higher than the threshold of a related standard request.

In summary, compared with a conventional insulation system consisting of mineral oil and Kraft paper, the FR3 impregnated T910 insulation system meets overloading transformer needs, which will improve the reliability of distribution transformers.

Author Contributions: Conceptualization, Y.X.; data curation, L.R.; funding acquisition, X.L.; investigation, X.Z., L.R., and H.Y.; methodology, X.Z., L.R., Y.X., and B.H.; project administration, Y.X. and X.L.; supervision, Y.X. and Q.L.; validation, B.H.; visualization, H.Y.; writing—original draft, X.Z., L.R., and H.Y.; writing—review and editing, X.Z., L.R., H.Y., and Y.X.

Funding: This paper is funded by China Southern Power Grid with the project No. GDKJQQ20152010.

Acknowledgments: The authors particularly wish to thank Xiaogang Xu for his support on the operation status of the distribution transformers and Sheng Liang for support on the thermal calculation of distribution transformers.

Conflicts of Interest: The authors declare no conflict of interest.

References

1. State Grid Corporation of China. *Technical Guide for High Overload Capacity Distribution Transformer of Rural Power Network*; Q/GDW 11190:2014; China Electric Power Press: Beijing, China, 2015.
2. China Southern Power Grid. *Technical Guidance of 10 kV Over-Loading Fluid Filled Transformer*; China Southern Power Grid: Guangzhou, China, 2015.
3. Cigre Working Group TF D1.01.10. *Aging of Cellulose in Mineral-Oil Insulated Transformers*; International Council on Large Electric Systems: Paris, France, 2007.
4. Frimpong, G.K.; Oommen, T.V. A survey of aging characteristics of cellulose insulation in natural ester and mineral oil. *IEEE Electr. Insul. Mag.* **2011**, *27*, 36–48. [CrossRef]
5. Yang, L.; Liao, R.; Sun, C. Influence of vegetable oil on the thermal aging rate of Kraft paper and its mechanism. In Proceedings of the 2010 International Conference on High Voltage Engineering and Application (ICHVE), New Orleans, LA, USA, 11–14 October 2010.
6. McShane, C.P.; Rapp, K.J.; Corkran, J.L. Aging of paper insulation in natural ester dielectric fluid. In Proceedings of the 2001 IEEE/PES Transmission and Distribution Conference and Exposition, Atlanta, GA, USA, 2 November 2001.
7. Lee, K.; Szewczyk, R.; Zhou, Y. A new insulation system for liquid-immersed distribution transformers. In Proceedings of the 5th European Conference on HV & MV Substation Equipments Power Utilities (MATPOST 2015), Lyon, France, 24–25 November 2015.
8. Zhang, X.; Shen, S.; Xu, Y. Thermal Evaluation of High-Temperature Insulation System for liquid-immersed transformer. In Proceedings of the IEEE Electrical Insulation Conference (EIC 2016), Montreal, QC, Canada, 19–22 June 2016.
9. IEEE C57.100 Working Group. *IEEE Standard Test Procedure for Thermal Evaluation of Insulation Systems for Liquid-Immersed Distribution and Power Transformers*; IEEE Std C57.100™-2011 (Revision of IEEE Std C57.100-1999); IEEE: New York, NY, USA, 2011.
10. International Electrotechnical Commission (IEC). *Electrical Insulation Systems (EIS)—Thermal Evaluation of Combined Liquid and Solid Components—Part 2: Simplified Test*; IEC TS 62332-2:2014; IEC: Geneva, Switzerland, 2014.
11. International Electrotechnical Commission (IEC). *Electrical Insulation Systems (EIS)—Thermal Evaluation of Combined Liquid and Solid Components—Part 1: General Requirements*; IEC TS 62332-1:2011; IEC: Geneva, Switzerland, 2011.

12. Wicks, R.C. Insulation system for liquid-immersed transformers—New materials require new methods for evaluation. In Proceedings of the IEEE Electrical Insulation Conference (EIC 2009), Montreal, QC, Canada, 31 May–3 June 2009.

13. Xu, Y.; Qian, S. Oxidation Stability Assessment of a Vegetable Transformer Oil under Thermal Aging. *IEEE Trans. Dielectr. Electr. Insul.* **2014**, *21*, 683–692. [CrossRef]

14. Lawson, W.G.; Simmons, M.A. Thermal Aging of Cellulose Paper Insulation. *IEEE Trans. Dielectr. Electr. Insul.* **1977**, *EI-12*, 61–66. [CrossRef]

15. Montsinger, V.M. Loading Transformers by Temperature. *Trans. Am. Inst. Electr. Eng.* **1930**, *49*, 776–790. [CrossRef]

16. American Society for Testing and Materials International (ASTM International). *Standard Test Method for Tensile Properties of Paper and Paperboard Using Constant-Rate-of-Elongation Apparatus*; ASTM D828-16; ASTM International: West Conshohocken, PA, USA, 2016.

17. International Electrotechnical Commission (IEC). *Electrical Insulation Materials—Thermal Endurance Properties—Part 3: Instructions for Calculating Thermal Endurance Characteristics*; IEC TS 60216-3:2009; IEC: Geneva, Switzerland, 2009.

18. International Electrotechnical Commission (IEC). *Electrical Insulation Materials—Thermal Endurance Properties—Part 2: Determination of Thermal Endurance Properties of Electrical Insulating Materials—Choice of Test Criteria*; IEC TS 60216-2:2005; IEC: Geneva, Switzerland, 2005.

19. Shroff, D.H.; Stannett, A.W. A Review of Paper Aging in Power Transformers. In *IEE Proceedings C—Generation, Transmission and Distribution*; IET: Washington, DC, USA, 1985; Volume 132, pp. 312–319.

20. Sierota, A.; Rungis, J. Electrical Insulating Oils Part I: Characterization and Pre-treatment of New Transformer Oils. *IEEE Electr. Insul. Mag.* **1995**, *11*, 8–20. [CrossRef]

21. Liao, R.; Liang, S.; Yang, L. The Improvement of Resisting Thermal Aging Performance for Ester-immersed Paper Insulation and Study on Its Reason. In Proceedings of the 2008 Annual Report Conference on Electrical Insulation Dielectric Phenomena, Quebec, QC, Canada, 26–29 October 2008.

22. Rapp, K.J.; McShane, C.P.; Luksich, J. Interaction Mechanisms of Natural Ester Dielectric Fluid and Kraft Paper. In Proceedings of the IEEE 15th International Conference on Dielectric Liquids (ICDL), Coimbra, Portugal, 26 June–1 July 2005.

23. Arroyo, O.H.; Jalbert, J. Temperature dependence of methanol and the tensile strength of insulation paper: Kinetics of the changes of mechanical properties during ageing. *Cellulose* **2017**, *24*, 1031–1039. [CrossRef]

24. Cigre Working Group WG A2.35. *Experience in Service with New Insulating Liquids*; International Council on Large Electric Systems: Paris, France, 2010.

25. McNutt, W.J.; Provost, R.L. Thermal life evaluation of high temperature insulation systems and hybrid insulation systems in mineral oil. *IEEE Trans. Power Deliv.* **1996**, *11*, 1391–1399. [CrossRef]

26. Lundgaard, L.E.; Hansen, W. Ageing of oil-impregnated paper in power transformers. *IEEE Trans. Power Deliv.* **2004**, *19*, 230–239. [CrossRef]

27. Lundgaard, L.E.; Hansen, W. Aging of Mineral Oil Impregnated Cellulose by Acid Catalysis. *IEEE Trans. Dielectr. Electr. Insul.* **2008**, *15*, 540–546. [CrossRef]

28. American Society for Testing and Materials International (ASTM International). *Standard Test Method for Dielectric Breakdown Voltage of Insulating Liquid Using VDE Electrodes*; ASTM D1816-12; ASTM International: West Conshohocken, PA, USA, 2012.

29. Tenbohlen, S.; Jovalekic, M.; Bates, L. Water Saturation Limits and Moisture Equilibrium Curves of Alternative Insulation Systems. In Proceedings of the CIGRE SC A2 & D1 Joint Colloquium 2011, Kyoto, Japan, 11–16 September 2011.

30. Wang, X.; Wang, Z.D. Particle Effect on Breakdown Voltage of Mineral and Ester Based Transformer Oils. In Proceedings of the Annual Report Conference on Electrical Insulation and Dielectric Phenomena (CEIDP 2008), Quebec, QC, Canada, 26–29 October 2008.

energies

MDPI

Article

Development of a Biodegradable Electro-Insulating Liquid and Its Subsequent Modification by Nanoparticles

Vaclav Mentlik, Pavel Trnka *, Jaroslav Hornak and Pavel Totzauer

Department of Technologies and Measurement, Faculty of Electrical Engineering, University of West Bohemia, 30614 Pilsen, Czech Republic; mentlik@ket.zcu.cz (V.M.); jhornak@ket.zcu.cz (J.H.); tocik@ket.zcu.cz (P.T.)
* Correspondence: pavel@ket.zcu.cz; Tel.: +420-377-634-518

Received: 23 January 2018; Accepted: 23 February 2018; Published: 27 February 2018

Abstract: The paper is focused on the possibility of replacing petroleum-based oils used as electro-insulating fluids in high voltage machinery. Based on ten years of study the candidate base oil for the central European region is rapeseed (Brassica napus) oil. Numerous studies on the elementary properties of pure natural esters have been published. An advantage of natural ester use is its easy biodegradability, tested according to OECD–301D (Organisation for Economic Co-operation and Development) standard, and compliance with sustainable development visions. A rapeseed oil base has been chosen for its better resistance to degradation in electric fields and its higher oxidation stability. The overall ester properties are not fully competitive with petroleum-based oils and therefore have to be improved. Percolation treatment and oxidation inhibition by a phenolic-type inhibitor is proposed and the resulting final properties are discussed. These resulting fluid properties are further improved using titanium dioxide (TiO_2) nanoparticles with a silica surface treatment. This fluid has properties suitable for use in sealed distribution transformers with the advantage of a lower price in comparison with other currently used biodegradable fluids.

Keywords: natural ester; titanium dioxide; resistivity; breakdown voltage; percolation treatment; antioxidant; surface treatment

---

1. Introduction

When designing a modern electrical machine, the technical aspects of the matter are not the only important ones—the environmental compatibility of the materials has become more important. Natural ester fluids have potential for industrial usage due to their good biodegradability (tested according to the OECD–301D standard [1]). A good biodegradable substance according to the standard is a substance which in the case of the contamination of soil, is naturally degraded by at least seventy percent by the soil bacteria in 28 days. This environmental property is an advantage in some technical solutions where devices require a liquid medium (for example, for cooling or to create a potential barrier to an electric field) for their operation and where there is a potential risk of soil contamination in the event of leakage [2]. Natural esters have this positive feature as they are examples of materials suitable for sustainable development applications. They can be produced with a competitive final price from local natural sources. These facts are the basic idea behind the preparation of a new electro-insulating liquid based on rapeseed oil. This liquid is expected to be competitive in its price and overall properties to the mineral-typically based electroinsulating liquids used in medium voltage transformers. There are commercial products already used in the transformers [3,4], however, still, several technical problems prevent them from being deployed. Well know are problems with their higher viscosity (a heat transfer issue) [5], high pour point [6], higher permittivity [7], low oxidation stability [8] (a hermetic design is required) and the high price of commercially produced esters, etc. This is also the reason why currently such fluids are deployed in Europe sporadically, mainly in headquarters or showrooms of international corporations or in specially protected areas. Less known are problems of virtually no legislative path allowing wider industrial use of natural esters, and durability problems in the terms of short high voltage impulse resistance [9,10], etc. Natural esters do also have advantages, e.g., lower degree of polymerization of cellulose compared to mineral fluids [11], high flash point [12] and inherent "moisture tolerance" [13].

2. Assessment of Natural Ester-Base Fluids for Further Development Based on Their Electrical Properties

In central Europe there are two widely spread representative natural esters that are suitable for industrial application—rapeseed (Brassica napus) [14] and sunflower (Helianthus annuus) [15] oil. Both crops are grown in several variations all over the world. The oil gathered from the seeds is a source material for several products, electro-insulating fluids included.

The composition of natural esters provides a very useful source of information about an oil's behavior. The structure of natural esters is based on a glycerol backbone with three attached fatty acid molecules. The nature of these fatty acids is decisive for the overall properties of the oil. For example, a high number of saturated fatty acids makes the oil stable to oxidation but also prone to solidification and high viscosity, whereas, a high number of poly- and mono-unsaturated fatty acids keeps the oil in a liquid state (low viscosity), even in lower temperatures, but makes it prone to oxidation. This is described in more detail in CIGRE Brochure N° 436 [16].

The first step is the assessment of a better base oil for further development. Experiments based on assessing the reaction and speed of the long-term thermal and electrical ageing of insulation systems containing these natural ester oils was done to provide data for a justified choice between rapeseed and sunflower oil. The main part of the experiment involved testing the liquids in the electrode system with a liquid (rapeseed and sunflower natural ester oil) and a solid (cellulose-based paper) insulation part simulating thus part of the power transformer. Each oil was inhibited with 0.4 wt % of phenolic inhibitor. The basic properties of the abovementioned fluids are provided in Table 1.

Table 1. Basic and required properties of untreated sunflower and rapeseed oil.

Parameter	Standard	Sunflower Oil *	Rapeseed Oil *	Unit
Density at 15 °C	ISO 3675, ISO 12185	0.92	0.917	$g·cm^{-3}$
Kinematic viscosity at 40 °C	ISO 3104	32.67	35.7	$mm^2·s^{-1}$
Dissipation factor at 90 °C	IEC 61620, IEC 60247	0.00202	0.0018	-
Relative permittivity	IEC 60247	3.93	3.44	-
Breakdown voltage	IEC 60156	63.77	61.54	kV/2.5 mm
Acid number	IEC 62021	0.042	0.092	mg KOH/g

* The values vary with the source of the oil—depending on the purity and water content.

The experiment itself proceeded as follows: samples of solid insulation (transformer board with thickness of 0.2 mm) were dried (24 h, at 80 °C), impregnated and then inserted between the electrodes in the electrode system shown in Figure 1. The oil itself was also dried using only heat (24 h, 90 °C to 80 ppm) before the beginning of the test. The electrode system consists of five electrodes with a diameter of 25 mm with rounded edges. The voltage levels for rapeseed oil were 4.3 to 7.4 kV, for sunflower oil 5.5 to 8.3 kV for similar time to breakdowns.

Figure 1. Electrode system for long-term electric stress.

Given the fact that oil behavior is strongly dependent on moisture [17], the whole test was performed under constant environment moisture monitoring, so the experiment itself was not affected. Table 2 shows the experimental results of the electric stress test. The comparison is made via the slope of the measured characteristics for each insulation system used, as the slope value serves as statistical indicator. The lower the slope value, the less prone to electrical stress the oil–paper system.

Table 2. Experiment results—slope comparison of measured characteristics.

Insulation System	Stress Type	Slope Value
Rapeseed oil + pressboard	AC 50 Hz	−0.884
Sunflower oil + pressboard	AC 50 Hz	−2.048

When an AC stress voltage is applied, the sunflower oil insulation system exhibits a higher rate of deterioration. The difference between the slope values (as shown in Table 2) is 1.16 which is statistically significant. From the measured characteristics (Figure 2) we can clearly state that the rapeseed oil has lower values of time to breakdowns, but the speed of degradation is lower.

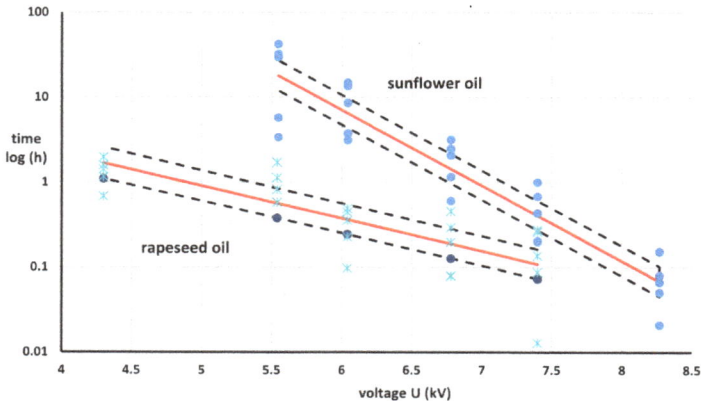

Figure 2. Time to breakdown characteristics for voltage stress exposition.

The verification of the system's thermal endurance was done by long-term thermal exposure at 140 °C for 1056 h and measurement of the dissipation factor and resistivity at regular intervals. Figures 3 and 4 show the values of the dissipation factor for rapeseed and sunflower oil, Figures 5 and 6 show the results of the volume resistivity for both oils.

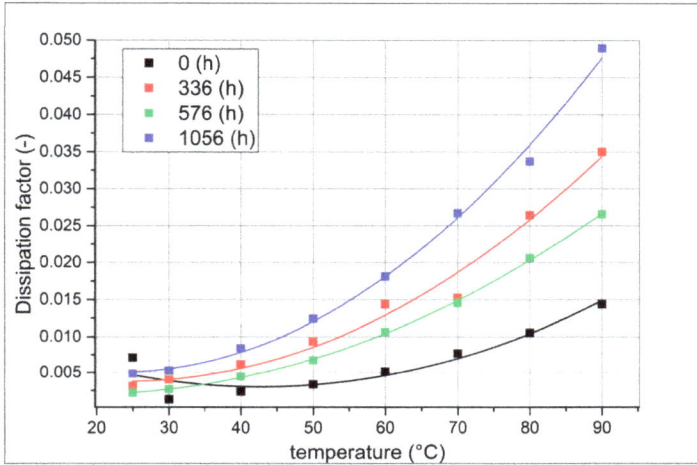

Figure 3. Thermal dependency of the dissipation factor of rapeseed oil at 140 °C exposure for different exposure times.

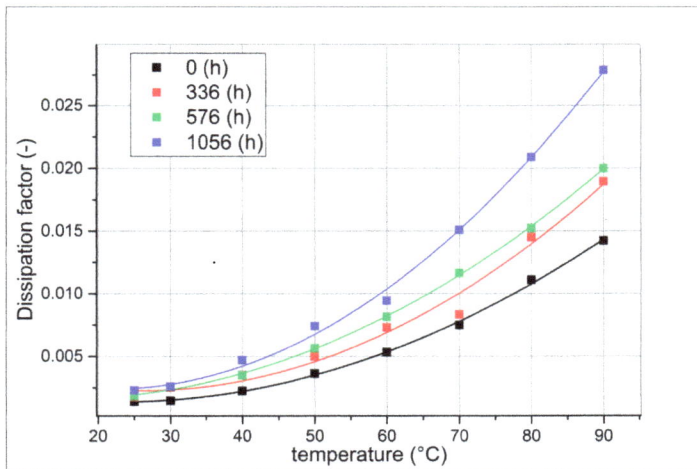

Figure 4. Thermal dependency of the dissipation factor of sunflower oil at 140 °C exposure for different exposure times.

The obtained thermal dependencies of the dissipation factor clearly states that the initial behavior of non-exposed oils is similar—their values are almost the same up to the value at 90 °C (a rise of 0.015). After thermal exposure (1056 h at 140 °C), the sunflower oil exhibits a slightly smaller change (tan δ at 90 °C 0.028) than rapeseed oil (tan δ at 90 °C 0.05).

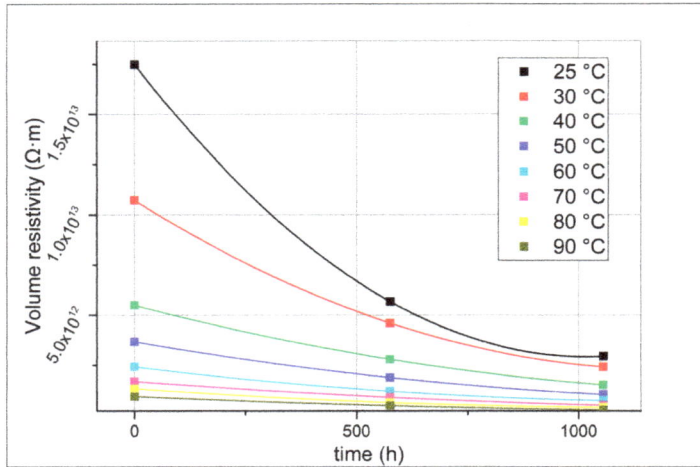

Figure 5. Time dependency of the resistivity of rapeseed oil at 140 °C exposure.

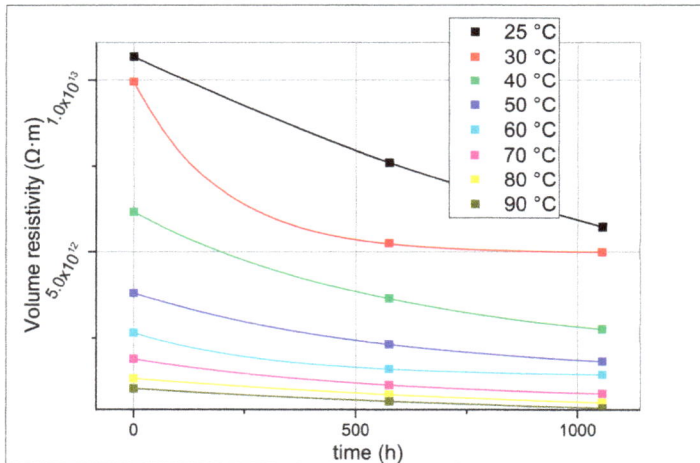

Figure 6. Time dependency of the resistivity of sunflower oil at 140 °C exposure.

The value of resistivity (25 °C) before thermal exposure at 140 °C is significantly better for rapeseed oil (1.75×10^{13} Ω·m) than sunflower oil (1.07×10^{13} Ω·m). The situation after thermal exposure (1056 h at 140 °C) changed a little. The value of resistivity (25 °C) is better for sunflower oil (5.71×10^{12} Ω·m vs. 9.45×10^{11} Ω·m for rapeseed oil), as the sunflower oil shows less change over the time of exposure. Finally, we can state that the behavior of both natural ester oils under elevated temperature is similar, with no conclusive differences.

When it comes to the overall classification of oils, the statistically significant difference of slope values under electrical stress comes to play. This, with better value of volume resistivity suggested rapeseed oil as the better of the two in our test.

3. Choosing the Rapeseed Base Oil as an Electro-Insulating Liquid

There are no fully implemented legislative guidelines for the usage of natural esters in transformers so far, but some properties are given in IEC 62770:2013 "Fluids for Electrotechnical Applications—Unused Natural Esters for Transformers and Similar Electrical Equipment" [18]. The values of the individual parameters are mentioned in the middle column of Tables 4 and 5.

Several commercially available rapeseed oil species have been verified, when searching for a raw material for the new electro-insulating liquid. It is obvious that these oils could not fully comply with the requirements of the Standard IEC 62770 without modifications. These crude oils cannot met the values of some the key parameters e.g., breakdown voltage (BDV), acid number, water content and the dissipation factor tan δ. Additionally, the values varied for different types and batches of oils. Therefore, it was necessary to adjust the insulating liquid for the desired purpose. In accordance with the present state of knowledge [19], this includes the technology of passing vegetable oil through a column filled with a suitable sorbent and the subsequent appropriate addition of an oxidation inhibitor at the corresponding concentration. This procedure was, of course, initially verified under laboratory conditions, where significant improvements in parameters were demonstrated to the extent that they now met the standard. The modified rapeseed oil passed the test of oxidative stability (TOS) according to method C in standard IEC 61125 [20]. The TOS had to be improved according to IEC-10/939/Q [21] from January 2014 to eliminate the deficiencies of the application of the standard EN 61125 to natural esters. The TOS, which were carried out showed the possibility of the application of modified vegetable oils as an electro-insulating liquid in accordance with the aforementioned standard IEC 62770. The TOS is based on the exposure of an oil sample by oxidation with the presence of copper, and eventually other materials. Changes in the measured parameters of the electro-insulating liquid are monitored before and after exposure. The rapeseed (not erucic) oil with low water content and a low level of volatile substances 0.02% (compared to other produced oils in food industry quality), produced by Usti Oils s.r.o, was chosen after laboratory tests as a base fluid for further development. Its fatty acid content is visible in Table 3.

Table 3. The components of rapeseed oil—used for the further preparation of an electro-insulating liquid.

Component	Unit	Value
Fatty acid content lower than C 16	%	0.08
Palmitic acid C 16:0	%	4.1
Stearic acid C 18:0	%	1.6
Oleic acid C 18:1	%	62.7
Linoleic acid C 18:2	%	19.1
Linolenic acid C 18:3	%	9.2
Erucic acid C 22:1	%	0.2

4. Preparation of the Electro-Insulating Ester Fluid

Based on the previous laboratory tests, a procedure for the treatment of a larger quantity of vegetable oil by a percolating device was designed (Figure 7a) and subsequently verified. The main scheme of the device is shown in Figure 7b. The abovementioned rapeseed oil flows from the reservoir (using a 200 L barrel) at a defined flow of 0.5 L/min through a sorbent column. The oil is subsequently returned to the container. The oil is heated to 60 °C for the viscosity reduction and passes through a composite pulp filter before returning to the tank. Two hundred liters of the oil is treated for approximately 24 h till the process is finished and requirements of the standard are fulfilled. The oil temperature of 60 °C has sufficiently adjusted viscosity to allow optimum flow and operation of the device. Figure 7 shows the cylindrical shape of the percolating column. The dimensions are: Height 70 cm, the diameter of circular base is 30 cm and the volume is approximately 50 L. The characteristic parameters of the original and modified rapeseed oil are shown in Table 4.

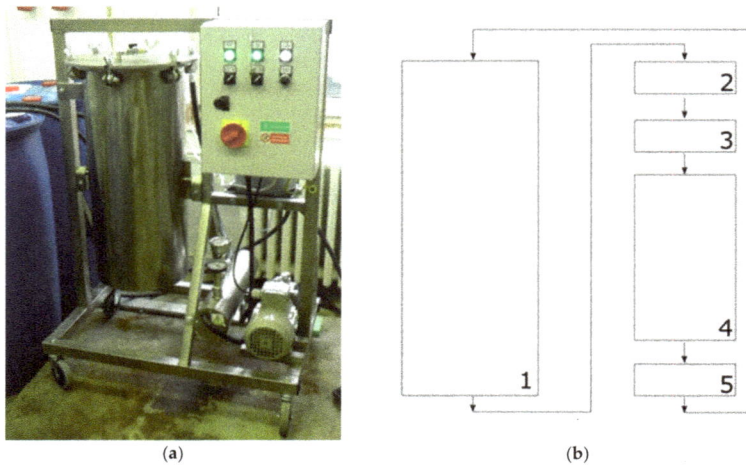

(a)   (b)

Figure 7. Percolation device: (**a**) Device for oil treatment; (**b**) scheme of rapeseed oil treatment (1—oil container, 2—pump, 3—heating unit, 4—sorbent column, 5—filter).

Table 4. Parameters of original and treated rapeseed oil.

Parameter (Unit)	Limit Value from IEC 62770	Original Oil	Modified Oil
Water content (mg/kg)	Max. 200	87.4	44.3
Density at 20 °C (g/mL)	Max. 1.0	0.913	0.915
Breakdown voltage (kV/2.5 mm)	Min. 35	73.5	58.5 *
Dissipation factor at 90 °C (-)	Max. 0.05	0.00477	0.00315
Acid number (mgKOH/g)	Max. 0.06	0.091	0.02

* No degassing.

As it is seen from Table 4, the acid number has been significantly lowered by the oil treatment and meets the criteria of IEC 62770. The value of breakdown voltage was reduced due to the partial foaming of oil during the treatment, but it is still within the limit value.

Another necessary step in the treatment of rapeseed oil is inhibition, i.e., the addition of an oxidation inhibitor. After an experimental verification, the anti-oxidation inhibitor dibutyl-para-cresol (DBPC) [22,23] was added to the obtained rapeseed oil. The results show that pure sunflower oil has slightly better oxidation stability than rapeseed, however after addition of phenolic type of antioxidant the stability of rapeseed oil is higher. The final concentration in the oil was set to 0.5 wt %.

The electro-insulating liquid from the rapeseed oil that has undergone the described modifications was ready for direct use in hermetic distribution transformers and it was protected by utility model number CZ 29982 [24] from 15. 11. 2016 with name ENVITRAFOL. The comparison of its properties with the values given in IEC 62770 is evident in Table 5.

Table 5. Comparison of the characteristic parameters of the ENVITRAFOL with the limit values according to IEC 62770.

Parameter (Unit)	Limit Value from IEC 62770	ENVITRAFOL
	Before Test of Oxidative Stability	
Appearance	Clear, free of sediment and suspension	Fulfill
Viscosity at 100 °C (mm^2/s)	Max. 15	8.26
Viscosity at 40 °C (mm^2/s)	Max. 50	35.84

Table 5. *Cont.*

Parameter (Unit)	Limit Value from IEC 62770	ENVITRAFOL
Pour point (°C)	Max. −10	−24
Water content (mg/g)	Max. 200	45.8
Density at 20 °C (g/mL)	1.0	0.915
Breakdown voltage (kV/2.5 mm)	Min. 35	60
Dissipation factor at 90 °C (-)	Max. 0.05	0.00358
Acid number (mgKOH/g)	Max. 0.06	0.011
Corrosive sulfur/DBDS	absent/below the limit of determination	absent
Additives antioxidants DBCP (wt %)	Max. 5	0.53
Additives all (wt %)	Max. 5	DBPC only
After Test of Oxidative Stability		
Dissipation factor at 90 °C (-)	Max. 0.5	0.02157
Viscosity at 40 °C (mm^2/s)	maximum increase of previous value of 30%	35.3
Acid number (mgKOH/g)	Max. 0.6	0.041

As can be seen from Table 5, the new application of natural esters resulting from the treatment of rapeseed oil fully complies with the requirements of IEC 62770. Its properties are comparable to similar products based on natural esters—Midel eN [3] or Envirotemp FR3 [4]. The main advantage is the usage of domestic raw materials with a relatively low price and suitable properties, one of which is biodegradability. This fact means that this oil is applicable in distribution transformers, many of which (in the Czech Republic alone there are more than 125,000 units) are often in exposed areas such as municipalities, and in proximity of drinking water sources or protected natural areas.

5. Improvements of Electrical Properties

As can be seen from the above, the new electro-insulating liquid resulting from the natural ester modification has suitable electrical properties. According to the present state of knowledge, the properties of the electro-insulating fluids can be further improved. The authors of numerous papers have dealt with the modification of the properties of mineral oils with nanofillers [25–31]. However, these fluids, as known, cannot meet the OECD–301D requirement for biodegradability. Therefore, the attention is focused on natural based electrical insulating liquids and their possible modification and improvement of some properties using inhibitors, depressants and nanoparticles.

The incorporation of the nanoparticles is proposed in order to improve the electrical properties, especially the key parameters—breakdown voltage and dissipation factor. Nanoparticles incorporated into the fluid system suppress the development of an electrical discharge in the liquid after applying the electrical potential.

The discharge path spreads by the nanoparticles, which leads to a slowing of its development and thus to an increase of the value of the flashover/breakdown voltage as well as to an improvement of the other electrical properties while preserving the overall physical and chemical parameters of the oil. Therefore the attention is focused on the modification of the natural ester by nanoparticles in this matter. The access is described in the following.

Based on the previous experience, and taking into consideration the conclusions of the above cited articles concerning mineral oils, these were the selected nanoparticles for the verification of the possibility of modifying the electrical properties of natural esters: TiO_2, Al_2O_3, SiO_2, ZnO. Table 6 gives an overview of the selected nanoparticles.

Table 6. Overview of selected nanoparticles to modify new oil properties.

Nanofiller Oxide	Purity (%)	Primary Particle Size (nm)	Variation	Surface Treatment
TiO_2	>96 TiO_2 <4 SiO_2	20	hydrophilic UV resistive	SiO_2
TiO_2	99+	20	-	-
Al_2O_3-γ	99.97	20–30	-	-

Table 6. *Cont.*

Nanofiller Oxide	Purity (%)	Primary Particle Size (nm)	Variation	Surface Treatment
Al₂O₃-γ	99.99	10	-	-
SiO₂	≥98 SiO₂ ≤2 (3-Aminopropyl) triethoxysilan	20	hydrophilic lipophilic	(3-Aminopropyl) triethoxysilan
SiO₂	99+	20	hydrophilic	-
ZnO	≥98 ZnO ≤2 (3-Aminopropyl) triethoxysilan	30	hydrophilic lipophilic	(3-Aminopropyl) triethoxysilan
ZnO	99+	20	-	-
ZnO	99+	30	-	-

The selection was carried out in several respects. The first is a type of nanoparticle. The second is material purity and particle size in nm. The third is surface treatment (ST), since it is known to play a significant role [29], particularly in terms of the incorporation of particles into the fluid system.

6. Incorporation of Individual Types of Nanoparticles

The first test selected for verification of prepared ENVITRAFOL nanofluids was the breakdown voltage test. This test was performed according to IEC 60156:1995 [32] in a standardized electrode system and ambient temperature, 2.5 mm distance, 440 mL of nanofluid, six breakdowns with five-minute pauses between breakdowns.

The weight content of nanofillers in the base oil was chosen from zero (pure ENVITRAFOL) and then fractions of a weight percent—0.05, 0.1, 0.2, 0.25, 0.3, 0.4, 0.5 and 1 wt %. All samples were prepared using the same procedure. The obtained results of maximal breakdown voltage are shown in Table 7.

Table 7. Maximal breakdown voltages of the nanofluids.

#	Nanofiller	Surface Treatment	Primary Particle Size (nm)	Max. Breakdown Voltage BDV (kV/2.5 mm)	Weight Content at max. BDV (%)
1	TiO₂	SiO₂	20	80.1	0.25
2	TiO₂	-	20	75.6	0.25
3	Al₂O₃-γ	-	10	74.3	0.2
4	Al₂O₃-γ	-	20–30	73.2	0.2
5	ZnO	-	20	72	0.3
6	ZnO	(3-Aminopropyl) triethoxysilan	30	70.8	0.3
7	ZnO	-	30	68.5	0.2
8	SiO₂	(3-Aminopropyl) triethoxysilan	20	65.9	0.2
9	SiO₂	-	30	64.5	0.2

The selected nanoparticles and ester oil samples were first dried at 110 °C for 18 h. During cooling the given amount of nanofillers was mixed into the oil. The resulting fluid system was then mixed with a double helix stirrer for one hour and subsequently a 200 W ultrasonic mixer was used for two hours using a frequency of 40 kHz. After this process, the sample of the fluid system was ready for testing. All nanoparticles listed in Table 6 have been tested. The obtained results are shown in Table 8 and Figure 8.

7. Results and Discussion

From the results (Table 7 and Figure 8), the following conclusions can be drawn: titanium dioxide with a silica surface treatment (SFT TiO₂) has the maximal positive effect on BDV (80.1 kV/2.5 mm by 0.25 wt % vs. 60 kV/2.5 mm for the pure oil) from tested samples. As expected, the influence of surface treatment of the nanoparticles plays significant role as seen from results of BDV of nanofluids with TiO₂ 75.6 kV/2.5 mm. As seen from results with Al₂O₃–γ: the values of BDV with the nanoparticles with a size of 10–20 nm are nearly the same as for 10 nm, 74.3 vs. 73.2 kV/2.5 mm. Thus, in this case, the particle size does not have an impact on BDV. The maximal BDV is in this case for 0.2% a mass of $Al_2O_3-\gamma$, but is still nearly 6 kV lower than for those with silica-coated TiO_2. BDV of nanofluids with ZnO is 72 (20 nm) respectively 70.5 kV/2.5 mm for SFT ZnO (30 nm) and 68.5 for ZnO (30 nm). In this case, nanofluids have the highest BDV with 20 nm nanoparticles, the surface treatment of the particles is beneficial in the case of 30 nm particles. The lowest BDV has nanofluid with SiO_2 nanoparticles. Here, the influence of surface treatment (3-aminopropyl) triethoxysilane is low (65.9 vs. 64.5 kV/2.5 mm). However the all samples proved increasing of BDV compared to pure ester oil at around 0.25 wt % of nanofiller.

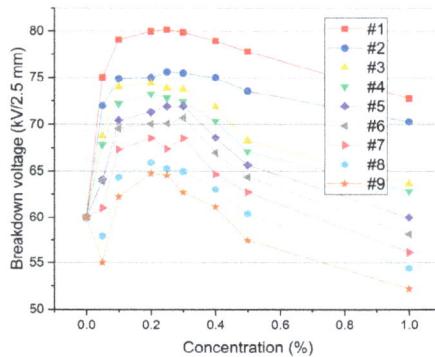

Figure 8. Breakdown voltages of nanofluids.

The increase of the BDV and volume resistivity by addition of nanofiller in concentration of 0.25 wt % can be explained by the creation of surface charges on the particles. These charges act against the outer electric field in the fluid, and thus the local field is lowered and electron cascades from the electrode are shifted toward a higher voltage. This phenomenon appears as an improvement of the listed properties. We can observe similar phenomena as e.g., in case of the electrical tree growing in material with and without nanoparticles. Reference [33] claimed that nanoparticles distributed in polymers could create a "fence effect", and make the growing path of the electrical tree twist and turn like a "Z" shape, so that more energy would be consumed. Similar phenomena can be expected in liquid dielectric, but due to the higher mobility of the molecules there is lower effect of the carbonization of the channel and therefore the increase of the high field at the end of the channel/avalanche does not occur. Still, due to the presence of nanoparticles, the more complicated path of the pre-breakdown channel consumes more energy and leads to the improvement of the measured BDV.

The abovementioned results tell us that the addition of nanoparticles significantly improved the BDV, which was already confirmed earlier for mineral oils [34]. Generally, the improvement of breakdown voltage of nanofluids may be caused by increased shallow trap density in the liquid. These traps could convert fast electrons to slow electrons by the trapping and de-trapping processes [35]. However, other factors are also important. The hydrophilicity of nanoparticles plays a great role in comparison with the effect of the surface treatment itself [36], because the hydrophilic surface can bind the water absorbed in the insulation liquid. Nevertheless, from the results it is also visible that the origin and usability of surfactant is also important [37]. The next one is the filler concentration. Figure 8 further shows a tendency of the BDV with the oil filling concentration. The BDV increases to a concentration of approximately 0.2 wt %, the maximum is in the range of 0.2–0.3 wt %. Higher concentrations of nanoparticles have a negative effect on the BDV. A higher concentration may cause an overlap of the interaction layers on the particle surface. Taking into account the fact that the particle is usually covered by 5–10 H_2O molecules tightly bound to its surface and that they may absorb additional water from oil, then the conductive paths may be created [38]. However, there are many additional variables which may also contribute in a minor effect to changes of breakdown voltage behavior (size, specific surface area, morphology, viscosity, moisture, etc.) [39]. The abovementioned investigations confirm that these hypotheses which were obtained for mineral oils are valid for natural ester oils as well.

The behavior of the modified oil samples is also interesting to observe in terms of statistical parameters, which are detailed in Table 8. The analysis of the basic statistical values of the BDV of the oils samples provides the titanium dioxide with silica surface treatment at 0.25 wt % at 25 °C has a minimal variation coefficient.

Table 8. Breakdown voltages (BDV) of the nanofluid samples.

#	1	2	3	4	5	6	7	8	9
Max (kV)	82.96	78.52	77.21	75.69	74.23	73.35	70.88	70.86	68.31
Min (kV)	77.61	73.11	69.46	69.33	69.15	66.06	61.96	60.16	59.52
X0.25 (kV)	79.12	73.83	73.03	71.66	70.18	67.56	66.15	62.42	63.36
X0.75 (kV)	81.03	76.68	76.31	75.24	73.12	72.31	69.62	68.64	66.62
σ^2 (kV²)—dispersion	3.24	3.81	7.53	5.59	3.70	7.94	9.78	15.63	9.04
v (%)—variation coefficient	2.25	2.58	3.72	3.25	2.67	4.02	4.64	6.06	4.66

The sample with the addition of surface treated with TiO_2 was further studied in order to characterize the basic electrical characteristics: Volume resistivity ϱ_v, minute polarization index PI and dissipation factor tan δ. The Tettex 2830/2831 Solid and Liquid Dielectric Analyzer with Tettex 2903 liquid dielectric capacitor (distance 2 mm, oil volume 40 mL) was used for these measurements.

In Figure 9, the temperature dependence of volume resistivity is presented. Volume resistivity (25 °C) of nanofluids with surface treated with TiO_2 particles rise from 1.09×10^{11} to 7.42×10^{11} $\Omega \cdot m$ for 0.25 wt %. The measured increase of the volume resistivity by addition of nanofillers is in agreement with the measured BDV increase. The particles create a charge on its surface (Stern layer [40]) acting against the main electric filed. The probability of electron scattering increases. This leads to a reduction of the impact energy of electrons and thus prevents the oil from ionizing [37]. The re-drop of the resistivity with a higher concentration than 0.25 wt % can be explained by an increase of free charge carriers caused by s higher concertation of the nanofillers and the activation of the interaction between particles and their interface layers. The column diagram in Figure 10 shows the dependence of the volume resistivity of the oil at 90 °C on the concentration of the surface treated TiO_2 with maximum at concentration of 0.25 wt %.

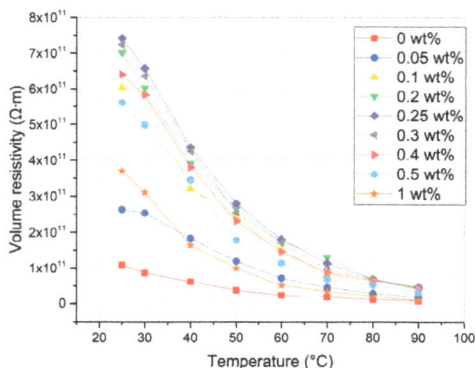

Figure 9. Temperature dependence of the volume resistivity of nanofluids with ST TiO_2.

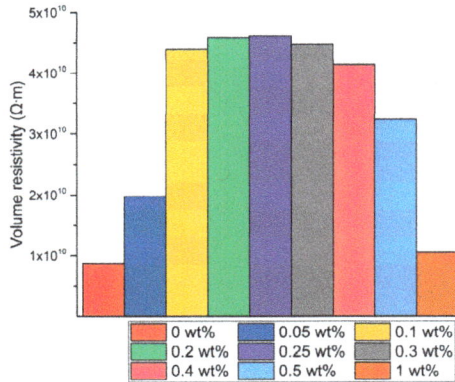

Figure 10. Dependence of the volume resistivity of nanofluids with ST TiO$_2$ on nanoparticle concentration, 90 °C.

The minute polarization index has the same trend since the same polarization current I$_{60s}$ is used for calculation. The application of the ST TiO$_2$ yielded an increase in the minute polarization index from 1.12 to 1.67. This fact corresponds with the phenomena of particle surface charge which is described above.

The dissipation factor tan δ was improved as well by the addition of ST TiO$_2$ (from 0.00148 to 0.00045), measured at 25 °C. The temperature dependence of the dissipation factor is presented in Figure 11. It can be seen that the minimum is at the concentration of 0.25 wt %. The dissipation factor increases for higher concentrations than 0.25 wt %. The column diagram in Figure 12 captures the dependence of the dissipation factor on the nanoparticle concentration at 25 ° C.

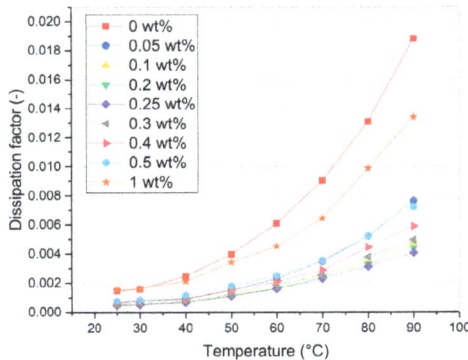

Figure 11. Temperature dependence of the dissipation factor of nanofluids with ST TiO$_2$.

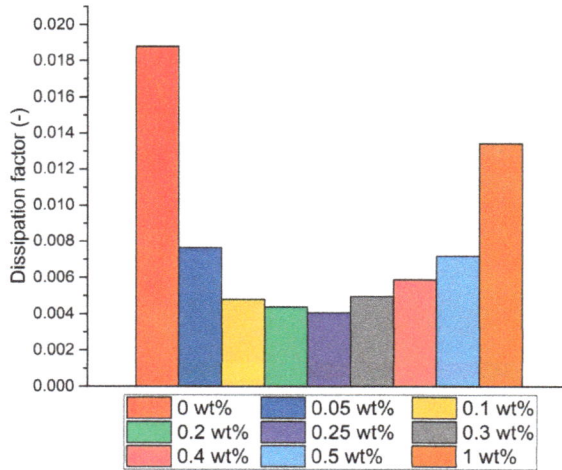

Figure 12. Content dependence of the dissipation factor of nanofluids with ST TiO$_2$, 25 °C.

As was mentioned above, the application of natural esters in transformers has several pros and cons [2,6,12]. The main advantage is their environmental compatibility, biodegradability and high flash (<200 °C) and fire points (<300 °C). For this reason, these natural ester oils can be classified as a "substance nonhazardous to waters" [41]. From the experiments which were carried out, it can be said that the presence of oxygen greatly affects the oxidative stability. For this reason, it is necessary to avoid the access of air to the transformer tank. This fact leads to the recommendation for application of natural ester oils in hermetically sealed transformers, in our case distribution transformers. The next discussed point is the higher viscosity of natural ester oil in comparison with mineral oil. This may cause an increase of temperature in the whole transformer [42]. This is solved by changing the cooling circuit design, which would lead to a reduction of the oil temperature [43]. On the other hand, it results in an increase of the power consumption of the oil pumps because these pumps must be able to pump a medium of higher viscosity [44]. The other thing which has to be changed is impregnation process of cellulose pressboard. This is also associated with a higher viscosity of natural ester oils. Increasing the impregnation temperature and prolonging the standing time will ensure satisfactory impregnation [45]. From these overviews is visible that it is difficult to apply natural esters without further transformer adjustments. However, if all these features are already included in the machine design itself, the potential of natural esters can be fully utilized e.g., in areas of increased environmental protection [6], such as shopping centers, protected natural areas, natural parks, proximity to water sources, etc. The next possibility for using a natural ester is retrofilling mineral oil transformers. It may reduce the moister content absorbed in cellulose-based pressboard due to moister migration and subsequent hydrolysis [13]. It leads to a slower aging process of cellulose pressboard [46].

8. Conclusions

The main aim of this paper was to describe the step-by-step development of an alternative electro-insulating liquid based on natural esters grown in Central Europe. The oxidative stability and long-term AC voltage exposure were the criteria parameters for rapeseed oil selection in comparison with sunflower oil which was also pre-tested. Rapeseed oil contains a high number of saturated fatty acids which make the oil oxidation stable and thus usable in long-term applications.

The original rapeseed oil had to be modified to meet the standards given by IEC 62770:2013. The percolation device was constructed. The acid number was significantly lowered by the oil treatment and acid number met the criteria given by IEC 62770:2013. The treated oil was also inhibited by the addition of DBPC in a concentration of 0.5 wt %. The modified rapeseed oil is ready for direct use in hermetized distribution transformers and it is protected by utility model number CZ 29982 with name ENVITRAFOL.

ENVITRAFOL was then modified with different kinds of nanoparticles to achieve even better parameters, according to the modern development trend. The silica-coated titanium dioxide (ST TiO$_2$) particles in a verified content of 0.25 wt % uniformly dispersed in ENVITRAFOL resulting in the largest improvement of several properties. The most significant change was observed in BDV behavior. The average value of BDV was increased from 60 to 80.15 kV/2.5 mm. However, other parameters were also improved. The dissipation factor decreased by an order, volume resistivity raised from 1.09×10^{11} to 7.42×10^{11}, 25 °C and the polarization index was raised from 1.12 to 1.67.

The resulting ENVITRAFOL with ST TiO$_2$ is proposed to be a liquid component of electrical isolation systems for electric equipment, e.g., distribution transformers (voltage levels tens of kV, rated power tens of kVA to tens of MVA) operated mainly in areas requiring increased environmental protection (municipalities, protected natural areas, natural parks, proximity to water sources, etc.) where any possible contamination would be an environmental issue. The use of this biodegradable electro-insulating nanofluid further supports the fact that its production is not technologically demanding and as an advantage it uses an easily accessible domestic renewable source which makes it economically beneficial compared to synthetic fluids.

Acknowledgments: This work is supported by the Ministry of Education, Youth and Sports of the Czech Republic under the RICE—New Technologies and Concepts for Smart Industrial Systems, project No. LO1607 and by the Student Grant Agency of the West Bohemia University in Pilsen, grant No. SGS-2018-016 Diagnostics and materials in electrotechnics.

Author Contributions: Vaclav Mentlik conceived and designed the experiments; Vaclav Mentlik, Pavel Trnka, Jaroslav Hornak, Pavel Totzauer performed the experiments, analyzed the data and wrote the paper. Jiri Ulrych a Jan Kubes performed some measurements.

Conflicts of Interest: The authors declare no conflict of interest.

References

1. OECD. *Ready Biodegradability*; Organisation for Economic Co-operation and Development: Paris, France, 1992.
2. Rafiq, M.; Lv, Y.Z.; Zhou, Y.; Ma, K.B.; Wang, W.; Li, C.R.; Wang, Q. Use of vegetable oils as transformer oils—A review. *Renew. Sust. Energy Rev.* **2015**, *52*, 308–324. [CrossRef]
3. MIDEL | Natural Ester Fluid/Vegetable Oil. Available online: www.midel.com/productsmidel/midel-en (accessed on 4 January 2018).
4. FR3 Fluid | Dielectric Ester Fluid | Mineral Oil Replacement | Cargill. Available online: www.cargill.com/bioindustrial/envirotemp/fr3 (accessed on 4 January 2018).
5. Perrier, C.; Beroual, A. Experimental investigations on mineral and ester oils for power transformers. In Proceedings of the 2008 IEEE International Symposium on Electrical Insulation, Vancouver, BC, Canada, 9–12 June 2008; pp. 178–181. [CrossRef]
6. Mehta, D.R.; Kundu, P.; Chowdhury, A.; Lakhiani, V.K.; Jhala, A.S. A review on critical evaluation of natural ester vis-a-vis mineral oil insulating liquid for use in transformers: Part 1. *IEEE Trans. Dielectr. Electr. Insul.* **2016**, *23*, 873–880. [CrossRef]
7. Guo, P.; Liao, R.; Hao, J.; Ma, Z.; Yang, L. Research on the temperature dielectric spectrum of vegetable oil, mineral oil and their relevant oil-impregnated papers. In Proceedings of the 2012 International Conference on High Voltage Engineering and Application, Shanghai, China, 17–20 September 2012; pp. 562–565. [CrossRef]
8. Bakrutheem, M.; Iruthayarajan, W.M.; Kumar, S.S. Investigation on the properties of natural esters blended with mineral oil and pyrolysis oil as liquid insulation for high voltage transformers. In *Intelligent and Efficient Electrical Systems*; Bhuvaneswari, M.C., Saxena, J., Eds.; Springer: Singapore, 2018; pp. 187–196. ISBN 978-981-10-4851-7.
9. Choi, S.; Huh, C. The lightning impulse properties and breakdown voltage of natural ester fluids near the pour point. *J. Electr. Eng. Technol.* **2013**, *8*, 524–529. [CrossRef]

10. Rapp, K.J.; Corkran, J.; McShane, C.P.; Prevost, T.A. Lightning impulse testing of natural ester fluid gaps and insulation interfaces. *IEEE Trans. Dielectr. Electr. Insul.* **2009**, *16*, 1595–1603. [CrossRef]

11. Martins, M.A.G. Vegetable oils, an alternative to mineral oil for power transformers- experimental study of paper aging in vegetable oil versus mineral oil. *IEEE Electr. Insul. Mag.* **2010**, *26*, 7–13. [CrossRef]

12. Bertrand, Y.; Lauzevis, P. Development of a low viscosity insulating liquid based on natural esters for distribution transformers. In Proceedings of the 22nd International Conference and Exhibition on Electricity Distribution, Stockholm, Sweden, 10–13 June 2013. [CrossRef]

13. Moore, S.; Rapp, K.; Baldyga, R. Transformer insulation dry out as a result of retrofilling with natural ester fluid. In Proceedings of the 2012 IEEE PES Transmission and Distribution Conference and Exposition (T&D), Orlando, FL, USA, 7–10 May 2012. [CrossRef]

14. Jahreis, G.; Schäfer, U. Rapeseed (*Brassica napus*) Oil and its Benefits for Human Health. In *Nuts and Seeds in Health and Disease Prevention*; Preedy, V., Watson, R.R., Patel, V., Eds.; Academic Press: Burlington, MA, USA, 2011; pp. 967–974. ISBN 978-0-12-375688-6.

15. Pal, D. Sunflower (*Helianthus annuus* L.) Seeds in Health and Nutrition. In *Nuts and Seeds in Health and Disease Prevention*; Preedy, V., Watson, R.R., Patel, V., Eds.; Academic Press: Burlington, MA, USA, 2011; pp. 967–974. ISBN 978-0-12-375688-6.

16. CIGRE. *Experiences in Service with New Insulating Liquids*; CIGRE Brochure N° 436; CIGRE: Paris, France, 2010.

17. Julliard, Y.; Badent, R.; Schwab, A.J. Influence of water content on breakdown behavior of transformer oil. In Proceedings of the Conference on Electrical Insulation and Dielectric Phenomena, Kitchener, ON, Canada, 14–17 October 2001; pp. 544–547. [CrossRef]

18. IEC. *Fluids for Electrotechnical Applications—Unused Natural Esters for Transformers and Similar Electrical Equipment*; IEC 62770:2013; International Electrotechnical Commission: Geneva, Switzerland, 2013.

19. Wilhelm, H.M.; Stocco, G.B.; Batista, S.G. Reclaiming of in-service natural ester-based insulating fluids. *IEEE Trans. Dielectr. Electr. Insul.* **2013**, *20*, 128–134. [CrossRef]

20. IEC. *Unused Hydrocarbon Based Isulating Liquids—Test Methods for Evaluating the Oxidation Stability*; IEC 61125:1992; International Electrotechnical Commission: Geneva, Switzerland, 1992.

21. IEC. *Proposed Revision of IEC 61125 Ed.1, Unused Hydrocarbon Based Insulating Liquids—Test Methods for Evaluating the Oxidation Stability*; IEC-10/939/Q; International Electrotechnical Commission: Geneva, Switzerland, 2014.

22. Kumar, S.; Iruthayarajan, M.W.; Bakrutheen, M.; Kannan, S.G. Effect of antioxidants on critical properties of natural esters for liquid insulations. *IEEE Trans. Dielectr. Electr. Insul.* **2016**, *23*, 2068–2078. [CrossRef]

23. Fofana, I. 50 years in the development of insulating liquids. *IEEE Electr. Insul. Mag.* **2013**, *29*, 13–25. [CrossRef]

24. Brázdil, J.; Černý, J.; Mentlík, V.; Trnka, P.; Košanová, L.; Kužílek, V. *Biodegradable Electrical Insulating Fluid*; Utility Model CZ 29 982; Industrial Property Office: Prague, Czech Republic, 2016.

25. Bin, D.; Li, J.; Wang, B.; Zhang, Z.T. Preparation and breakdown strength of Fe_3O_4 nanofluid based on transformer oil. In Proceedings of the 2012 International Conference on High Voltage Engineering and Application, Shanghai, China, 17–20 September 2012; pp. 311–313. [CrossRef]

26. Mergos, J.A.; Athanassopoulou, M.D.; Argyropoulos, T.G.; Dervos, C.T. Dielectric properties of nanopowder dispersions in paraffin oil. *IEEE Trans. Dielectr. Electr. Insul.* **2012**, *19*, 1502–1507. [CrossRef]

27. Das, S.K.; Choi, S.U.; Yu, W.; Pradeep, T. *Nanofluids: Science and Technology*; John Wiley & Sons: Hoboken, NJ, USA, 2008; ISBN 13 978-0470074732.

28. Fuxin, W.; Ming, D.; Jianzhuo, D.; Ming, R.; Rixin, Y. Study of Breakdown Mechanism of Transformer Oil Based on ZnO Nanoparticles. In Proceedings of the International Symposium on High Voltage Engineering 2015, Pilsen, Czech Republic, 23–28 August 2015; pp. 1–4.

29. Putra, N.; Roetzel, W.; Das, S.K. Natural Convection of Nano-fluids. *Heat Mass Transfer.* **2003**, *39*, 775–784. [CrossRef]

30. Sima, W.X.; Cao, X.F.; Yang, Q.; Song, H.; Shi, J. Preparation of Three Transformer Oil-Based Nanofluids and Comparison of Their Impulse Breakdown Characteristics. *Nanosci. Nanotechnol. Lett.* **2014**, *6*, 250–256. [CrossRef]

31. Lv, Y.; Rafiq, M.; Li, C.; Shan, B. Study of Dielectric Breakdown Performance of Transformer Oil Based Magnetic Nanofluids. *Energies* **2017**, *10*, 1025. [CrossRef]

32. IEC. *Insulating Liquids—Determination of the Breakdown Voltage at Power Frequency—Test Method*; IEC 60156:1995; International Electrotechnical Commission: Geneva, Switzerland, 1995.

33. Danikas, M.G.; Tanaka, T. Nanocomposites-a review of electrical treeing and breakdown. *IEEE Electr. Insul. Mag.* **2009**, *25*, 19–25. [CrossRef]

34. Jin, H.; Andritsch, T.; Tsekmes, I.A.; Kochetov, R.; Morshuis, P.H.F.; Smit, J.J. Properties of mineral oil based silica nanofluids. *IEEE Trans. Dielectr. Electr. Insul.* **2014**, *19*, 1100–1108. [CrossRef]

35. Du, Y.; Lv, Y.Z.; Li, C.; Chen, M.; Zhou, J.; Li, X.; Zhou, Y.; Tu, Y. Effect of electron shallow trap on breakdown performance of transformer oil-based nanofluids. *J. Appl. Phys.* **2011**, *110*. [CrossRef]

36. Jin, H.; Morshuis, P.H.F.; Smit, J.J.; Andritsch, T. The effect of surface treatment of silica nanoparticles on the breakdown strength of mineral oil. In Proceedings of the 2014 IEEE 18th International Conference on Dielectric Liquids, Bled, Slovenia, 29 June–3 July 2014; pp. 1–4. [CrossRef]

37. Du, Y.; Lv, Y.Z.; Wang, F.; Li, X.; Li, C. Effect of TiO_2 nanoparticles on the breakdown strength of transformer oil. In Proceedings of the 2010 IEEE International Symposium on Electrical Insulation, San Diego, CA, USA, 6–9 June 2010; pp. 1–3. [CrossRef]

38. Zou, C.; Fothergill, J.; Rowe, S. The effect of water absorption on the dielectric properties of epoxy nanocomposites. *IEEE Trans. Dielectr. Electr. Insul.* **2008**, *15*, 106–117. [CrossRef]

39. Rafiq, M.; Lv, Y.; Li, C. A review on properties, opportunities, and challenges of transformer oil-based nanofluids. *J. Nanomater.* **2016**, *2016*, 1–23. [CrossRef]

40. Lewis, T.J. Interfaces: Nanometric dielectrics. *J. Phys. D* **2005**, *38*, 202–212. [CrossRef]

41. Tenbohlen, S.; Koch, M. Aging Performance and moisture solubility of vegetable oils for power transformers. *IEEE Trans. Power Deliv.* **2010**, *25*, 825–830. [CrossRef]

42. Dombek, G.; Goscinski, P.; Nadolny, Z. Comparison of mineral oil and esters as cooling liquids in high voltage transformer in aspect of environment protection. *E3S Web Conf.* **2017**, *14*, 01053. [CrossRef]

43. Smith, J.S.; Beaster, B.L. Design and test experience with natural ester fluid for power transformers update. In Proceedings of the 2009 IEEE Power & Energy Society General Meeting, Calgary, AB, Canada, 26–30 July 2009. [CrossRef]

44. Fritsche, R.; Rimmele, U.; Schäfer, M. *Prototype 420 kV Power Transformer Using Natural Ester Dielectric Fluid*; Siemens, A.G.: Nuremberg, Germany, 2014.

45. Darwin, A.; Perrier, R.; Foliot, P. The use of natural ester fluids in transformer. In Proceedings of the MATPOST Conference, Lyon, France, 15–16 November 2007.

46. Bandara, K.; Ekanayake, C.; Saha, T.; Ma, H. Performance of Natural Ester as a Transformer Oil in Moisture-Rich Environments. *Energies* **2016**, *9*, 258. [CrossRef]

energies

MDPI

Article

Lightning Impulse Withstand of Natural Ester Liquid

Stephanie Haegele [1],*, Farzaneh Vahidi [1], Stefan Tenbohlen [1], Kevin J. Rapp [2] and Alan Sbravati [2]

[1] Institute of Power Transmission and High Voltage Technology, University of Stuttgart, 70569 Stuttgart, Germany; farzaneh.vahidi@ieh.uni-stuttgart.de (F.V.); stefan.tenbohlen@ieh.uni-stuttgart.de (S.T.)

[2] Cargill Inc., Cargill Industrial Specialties—Dielectric Fluids, Plymouth, MN 55441, USA; kevin_rapp@cargill.com (K.J.R.); alan_sbravati@cargill.com (A.S.)

* Correspondence: stephanie.haegele@ieh.uni-stuttgart.de; Tel.: +49-(0)-711/685-67858

Received: 7 July 2018; Accepted: 25 July 2018; Published: 28 July 2018

Abstract: Due to the low biodegradability of mineral oil, intense research is conducted to define alternative liquids with comparable dielectric properties. Natural ester liquids are an alternative in focus; they are used increasingly as insulating liquid in distribution and power transformers. The main advantages of natural ester liquids compared to mineral oil are their good biodegradability and mainly high flash and fire points providing better fire safety. The dielectric strength of natural ester liquids is comparable to conventional mineral oil for homogeneous field arrangements. However, many studies showed a reduced dielectric strength for highly inhomogeneous field arrangements. This study investigates at which degree of inhomogeneity differences in breakdown voltage between the two insulating liquids occur. Investigations use lightning impulses with different electrode arrangements representing different field inhomogeneity factors and different gap distances. To ensure comparisons with existing transformer geometries, investigations are application-oriented using a transformer conductor model, which is compared to other studies. Results show significant differences in breakdown voltage from an inhomogeneity factor of 0.1 (highly inhomogeneous field) depending on the gap distance. Larger electrode gaps provide a larger inhomogeneity at which differences in breakdown voltages occur.

Keywords: dielectric breakdown voltage; dielectric liquids; natural ester liquids vs. mineral oil; power transformers; vegetable oils

1. Introduction

Natural ester liquids (NE) are increasingly used as insulating liquids in distribution and power transformers. A reason for this is the environmental advantage in comparison to the traditionally used mineral oil (MO). Good biodegradability, generally high flash and fire points, low toxicity and their contribution to lower risks for humans and the environment has already been evaluated [1] for the natural ester liquid used in this contribution. The dielectric strength of insulating liquids is one of the main parameters relevant for power transformers. Several studies showed a comparable dielectric strength of the same used natural ester liquids compared to mineral oil under homogeneous and slightly inhomogeneous fields at lightning impulse (LI) [2–4] or with a different natural ester liquid [5].

Tests were performed with different polarities over a range of gap distances and electrode diameters. Other tests showed significantly lower dielectric strength of natural ester liquids at highly inhomogeneous fields using needle—plate arrangements under LI [6–8]. Differences in breakdown voltage occur due to differences in streamer propagation between (well-investigated) mineral oils [9–12] and natural ester liquids [13–16]. Differences increase by growing gap distances and by growing inhomogeneity factors. Insulating liquids used for the described tests are Envirotemp[TM] FR3[TM] (Minnetonka, MN, USA) fluid as natural ester liquid produced by Cargill, Inc. and Nytro Lyra X (Stockholm, Sweden) as mineral oil produced by Nynas Inc.

This study evaluates the breakdown voltages of arrangements with different electrode settings forming a large range of inhomogeneity factors. Its aim is to define those parameters at which differences in breakdown voltage and breakdown field strength emerge between different liquids.

To get closer to transformer geometry, a series of tests with a transformer conductor at defined gap distance and inhomogeneity were conducted to compare the results to the results of the initial study.

2. Measurement of Breakdown in Insulating Liquid

Two different setups were used. The used generator and measurement procedure are the same for all following investigations.

2.1. Lightning Impulse Generating and Measurement Setup

A 1 MV Marx generator was used for all tests. Its maximum rated energy is 30 kJ. A standard 1.2/50 μs lightning impulse was applied in all tests. Investigations with different inhomogeneity factors were performed at negative lightning impulse, investigations with the transformer conductor were conducted under positive lightning impulse in order to provide a complete, but feasible range of measurement results and to be able to compare to previous tests.

Inhomogeneity tests were performed with large (volume $V = 16$ L) and small ($V = 1.6$ L) measurement cells adapted to different voltage levels which limits the quantities of required insulating liquid. Transformer conductor tests were performed in a separate large steel tank (see Figure 1) with bushing (BIL 750 kV).

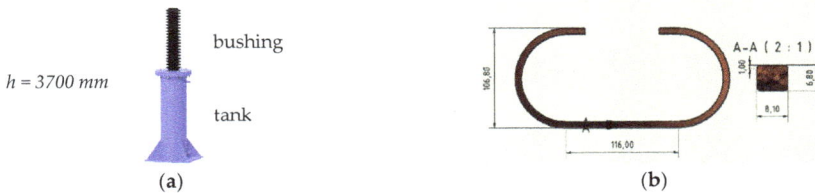

Figure 1. (**a**) Steel test tank ($V = 320$ L) with bushing and (**b**) copper electrode for transformer conductor tests.

Different electrode and gap arrangements were tested. Some configurations were tested in both test cells to exclude possible influences of volume effects. The used steel electrodes were mainly manufactured at the institute. Needle electrodes were purchased. The specified tip radius of the steel needles was confirmed by scanning electron microscopy (SEM).

The electrode diameters used were 120 μm, 0.6 mm, 0.8 mm, 3 mm and 13 mm. Gap distances used were 10, 15, 25, 40 and 50 mm. Using larger gap distances than 50 mm would have resulted in breakdown voltages too high to handle (oil volume, electrode abrasion) for these investigations for all configurations but the needle. As the focus of our investigations was on degrees of inhomogeneity between the highly inhomogeneous needle configuration and slightly inhomogeneous sphere configuration and not on the extremes, investigated gap distances were limited to 50 mm. An overview of the test configurations used is given in Table 1.

Electrodes used for transformer conductor tests were made from copper and use the same dimensions as the conductor used in former tests at the institute to guarantee comparability. Conductors were not wrapped with paper; bare conductors were used. The design was changed in comparison to former tests to avoid the possibility of a breakdown from the ends of the conductor to the rounded ground plate. The minimal radius of curvature of the conductor used for these tests was 1 mm.

Table 1. Configurations for inhomogeneity tests under lightning impulse (LI) at negative polarity.

10 mm	15 mm	25 mm	40 mm	50 mm
Sphere	Sphere	Sphere	-	-
Blunt point (Bp) 3 mm	Bp 3 mm	Bp 3 mm	-	-
Bp 0.8 mm	Bp 0.8 mm	Bp 0.8 mm	-	-
Bp 0.6 mm	Bp 0.6 mm	Bp 0.6 mm	-	-
Needle 120 µm	-	Needle 120 µm	Needle 120 µm	Needle 120 µm

2.2. Procedure

Oil samples were prepared from dried, degassed and filtered samples. Relative moisture *rH* was set to values smaller than 12% for all liquids and tests. Moisture measurement and breakdown tests were performed at ambient temperature. Filtering cartridges with element size smaller than 5 µm were used. Ten breakdown tests per configuration and per insulating liquid are performed for the sphere, blunt point and needle tests. Waiting times before starting tests were much longer for natural ester liquid than for mineral oil and depend on the oil volume. Minimum waiting time for tests with the 1.6 L cell was five minutes for mineral oil and 15 min for natural ester liquid. Liquid was poured in carefully; nevertheless, minimum waiting time was 15 min for mineral oil and 30 min for natural ester liquid for the large 16 L test cell. Necessary waiting time was investigated in five-minute steps previously. Too short waiting time resulted in lower breakdown voltage and especially higher standard deviation. Previous tests were conducted to define the waiting time at which no considerable decreases in breakdown voltage and standard deviation could be noticed compared to the five-minute shorter waiting time. Waiting time for tests with the large steel tank was one day after complete change of liquid and two hours in between tests. Thirty breakdown tests were performed for each insulating liquid at this arrangement.

2.3. Electrode Conditioning

As surface defects are the dominant breakdown effect during LI stress [17], electrode surface effects need to be considered and tested before the actual investigation. Preliminary tests were conducted to define electrode replacement intervals and polishing intervals [18].

A preliminary test was performed to assure the usability of the chosen needle electrodes. Needles were electrically tested at different stress levels: No stress, one impulse, 10 impulses, 15 impulses, one breakdown and several breakdowns. Results and scanning electron microscope images can be found in [18]. ASTM D3300 standard with which mentioned investigations were performed suggests electrode polishing intervals of five breakdowns for sphere electrodes and immediate replacement of needle electrodes after one breakdown. These suggestions were considered, and intervals were determined for the electrode configurations of blunt points with radii smaller than sphere and larger than needle tip. Blunt points were manufactured and polished by the institute's workshop.

The polishing of blunt points with small tip radii changes radii over time. Therefore, blunt points with 0.6 and 0.8 mm are polished only one time. After the second use they are replaced. Blunt points with larger radii could be polished several times. Corresponding polishing intervals can be found in [18].

Additionally, electrode conditioning is investigated. Blunt points are stressed with many breakdowns. In most cases, the first breakdown after polishing shows a significantly larger breakdown voltage than following impulses. Therefore, the first breakdown after polishing is not taken into account for statistics.

Preliminary tests were performed to determine the replacement intervals of the copper conductor and the influence of breakdown on the ground electrode (plane with rounded edges) for conductor tests.

2.4. Calculation of Inhomogeneity Degrees

In order to quantify the degree of inhomogeneity, the Schwaiger Factor η is used.

$$\eta = E_{mean}/E_{max} \tag{1}$$

The Schwaiger Factor defines the relation between mean (E_{mean}) and maximum (E_{max}) field strength and can be derived analytically from geometry factors for simple arrangements [4,19]. For arrangements that are more complex, computer-aided design (CAD) and numerical field simulation is necessary to calculate inhomogeneity factors. Field simulation is performed using a three-dimensional electrostatic model. Special care needs to be taken for meshing of needle arrangements (Figure 2). The maximum calculated field strength of a configuration depends on the selected mesh for the use of default physics-controlled mesh options. Element size needs to be reduced to the point where no dependency between varying element size and maximum field strength is given anymore.

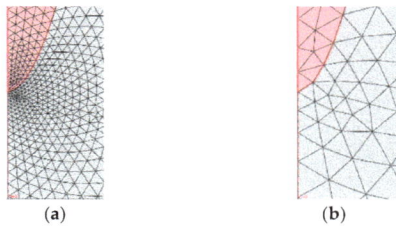

(a)   (b)

Figure 2. Meshing of needle electrode in field simulation. (**a**) manually with finer mesh than the finest automatic meshing; (**b**) automatic with "extra fine" mesh.

Considered Schwaiger Factors (see Figure 3 range between highly inhomogeneous arrangements commonly made up with needle electrodes at high voltage potential and sphere or plate electrodes at ground potential and only slightly inhomogeneous arrangements like sphere-to-sphere. The aim is to represent a large range of inhomogeneities at the limits of radii used in transformer design. The field inhomogeneity η for transformer conductor tests is calculated to be $\eta = 0.21$.

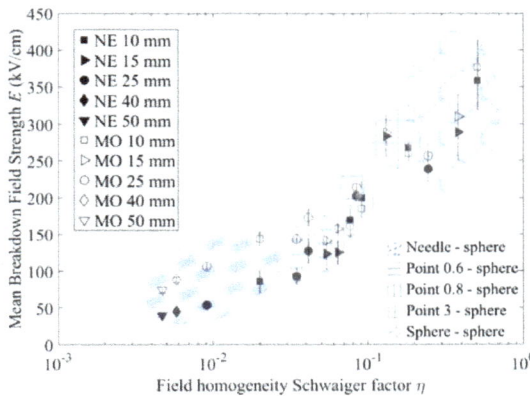

Figure 3. Mean breakdown field strength over the range of investigated inhomogeneity at negative polarity for mineral oil and natural ester liquid.

3. Measurement Results

Breakdown voltages and breakdown field strengths are determined for all investigated configurations. First, breakdown behavior of fresh insulating liquids is determined.

3.1. Mean Breakdown Field Strength, Mean Breakdown Voltage and Withstand Voltage

Figure 4 shows mean breakdown voltage versus gap distance; Figure 3 shows mean breakdown field strength over the investigated range of inhomogeneity. Field strengths during an ongoing discharge can be a lot higher than the calculated ones due to space charges depending on the distance of the streamer tip from the electrode and the diameter of the streamer. This effect was investigated in [20–25].

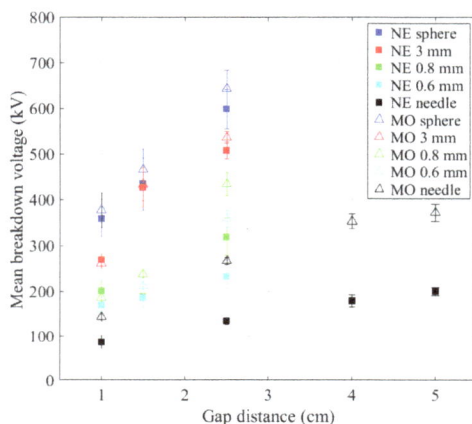

Figure 4. Mean breakdown voltages versus gap distance at negative polarity for mineral oil and natural ester liquid.

Mean breakdown voltages are comparable between the two investigated liquids for small gap distances and slightly inhomogeneous electrode arrangements. Natural ester liquid shows significantly reduced mean breakdown voltages for highly inhomogeneous arrangements at large gaps (blunt points 0.6 mm, 0.8 mm and needle at gaps ≥25 mm).

Mean breakdown field strengths are plotted versus Schwaiger factors to show the influence of field inhomogeneity. They are small for highly inhomogeneous electrode arrangements (needle) compared to slightly inhomogeneous electrode arrangements (sphere). An explanation therefore is that highly inhomogeneous arrangements show high breakdown field strength next to the inhomogeneous needle tip, but comparably small field strength in the main volume. Homogeneous or slightly inhomogeneous arrangements show a more homogeneous distribution of the field strength over the whole arrangement showing smaller maximum values than the highly inhomogeneous arrangements, but larger mean values. Overall, mean breakdown field values are decreasing from small inhomogeneities to large inhomogeneities with a few exemptions for the small gap distance tests with 3 mm blunt point.

Mineral oil shows higher mean breakdown field strength than natural ester liquid for the configurations with highly inhomogeneous fields from an inhomogeneity factor η of $\eta < 0.1$. Differences between the two liquids increase towards larger gap distances and higher inhomogeneity.

The needle configurations at 40 mm and 50 mm gap distance show the largest differences between the two liquids: Natural ester liquid depicts 60% of the mean field of mineral oil at 50 mm gap distance. At inhomogeneities of $0.1 < \eta < 1$, both insulating liquids show comparable mean breakdown field strengths for all electrode configurations and gap distances.

Regarding the mean breakdown voltages, the following effect can be observed: The larger the electrode gap and the larger the homogeneity of the field, the larger the resulting breakdown voltage. Natural ester liquids and mineral oils show comparable breakdown values for homogeneous and slightly inhomogeneous field strength values with $\eta > 0.1$. Mean breakdown voltages drop significantly for natural ester liquids for inhomogeneous arrangements. With larger gap distances, the differences between the breakdown voltages of natural ester liquids and mineral oil also increase in this area. The standard deviation is comparable for both insulating liquids for nearly all configurations. Configurations with low breakdown voltages generally show a low standard deviation. The more homogeneous the electrode arrangement, the higher the standard deviation. Breakdown data is fitted to normal and Weibull distribution. Normal distribution fitting is slightly better for the measured impulse breakdown data than Weibull distribution. Furthermore, 2% withstand voltages are calculated.

Slightly inhomogeneous configurations with higher standard deviation show lower withstand voltages than highly inhomogeneous arrangements with low standard deviation compared to the mean breakdown values. Figure 5 shows Weibull and normal distribution fitted to breakdown data for an exemplarily configuration.

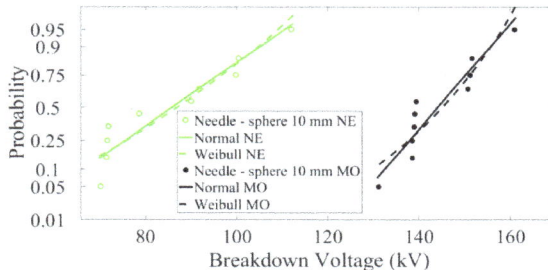

Figure 5. Weibull and normal distribution for natural ester liquid and mineral oil at 10 mm gap distance fitted to the measured needle—sphere data at negative polarity.

3.2. Homogeneity Factor and Breakdown Voltage—Differences between Mineral Oil and Natural Ester Liquid

A Schwaiger Factor $\eta_s = 0.1$ can be determined as the degree of inhomogeneity required to see differences between the two liquids. Considering the entire available range of gap and electrode configurations by extrapolating the general interrelationships of the measured data, it is expected that for far larger gap distances than the tested ones, η_s is shifted towards larger values of η. For gap distances around 10 mm, η_s is shifted towards smaller values of η. The selected range of η for further testing is determined to be $0.03 < \eta < 0.3$. Differences in breakdown behavior of natural ester liquids and mineral oils at highly inhomogeneous field condition are caused by different streamer propagation mechanisms. Easier propagation at high propagation modes leads to lower mean and withstand breakdown voltages. A possible explanation for this effect is the presence of polyaromatic molecules in mineral oil that do not appear in natural ester liquids [26].

3.3. Conductor Breakdown Test

A new test set-up is built to investigate breakdown voltages and breakdown field strengths of natural ester liquids compared to mineral oil in a set-up more suitable to electric field constellations occurring in a real transformer. The gap distance is set to 20 mm. Results are again fitted to normal and Weibull distribution to determine 1% withstand voltages. Results for all distributions are shown in Tables 2 and 3.

Table 2. Breakdown results for conductor test at positive polarity, normal distribution fitting for mineral oil (MO) and natural ester liquid (NE).

Liquid	Normal Distribution		
	$U_{d,1\%}$ (kV)	$U_{d,50\%}$ (kV)	σ
MO	291.6	356.6	28.0
NE	252.3	318.3	28.3

Table 3. Breakdown results for conductor test at positive polarity, Weibull distribution fitting.

Liquid	Weibull Distribution			
	$U_{d,1\%}$ (kV)	$U_{d,50\%}$ (kV)	α	β
MO	248.3	358.5	370.1	11.5
NE	222.8	321.2	331.5	11.6

Table 2 shows normal distribution fitted breakdown values with 50% breakdown and 1% withstand voltages for natural ester liquid and mineral oil. Natural ester liquids show 89% of the breakdown voltage of mineral oil for 50% mean values and 87% for 1% withstand voltages. The results for the Weibull distribution fitted data in Table 3 depict natural ester liquids holding 90% of the mean breakdown voltage values of mineral oil, and 90% for 1% withstand voltages. A comparison between the fittings of normal and Weibull distribution shows higher 50% breakdown voltages for Weibull distribution and higher 1% withstand voltages for normal distribution for both insulating liquids. A probability plot is given in Figure 6. Normal distribution fits slightly better than Weibull distribution to the measurement data for both insulating liquids.

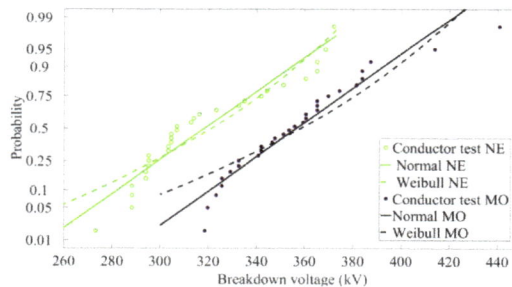

Figure 6. Weibull and normal distribution for natural ester liquid and mineral oil fitted to the measured conductor data at a gap distance of 20 mm and positive polarity.

The LogLikelihood method was used to determine the quality of each fit. The quality of the fits is comparable for both insulating liquids and distributions. Standard deviation of both insulating liquids is comparable and fulfills the standard requirements.

A comparison to other investigations with comparable electrode arrangement and comparable low relative moisture [2] is provided in Figure 7. The main difference between the two setups is a change in curvature of the ends of the transformer conductor. This change was performed to compare if the larger inhomogeneity represented by the exposed ends of the conductor in the former tests possibly influences the differences between natural ester liquid and mineral oil. The mentioned conductors are shown in Figures 1 and 8.

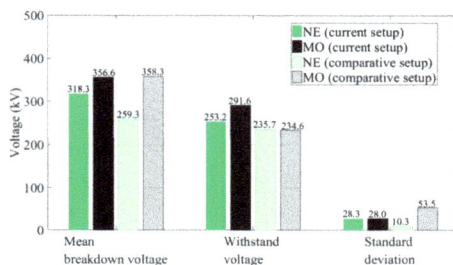

Figure 7. Comparison with results from the former measurement setup in [2] with comparable gap arrangement at positive polarity.

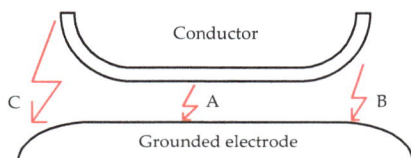

Figure 8. Curvature of conductors used in comparative tests in [2].

The experiments described here show reduced differences in breakdown voltage between natural ester liquids and mineral oil. The mean breakdown voltage of natural ester liquids is higher than in a previous study, compared to the mineral oil being equal. Comparative tests showed a significantly higher standard deviation for mineral oil than for natural ester liquids. The withstand voltages of both liquids are therefore comparable. Current experiments show comparable standard deviations for both insulating liquids. As it was also observed for the mean values, natural ester liquids show slightly lower withstand voltages than mineral oil (87%). Because the knowledge of comparative experiments is only based on the information given in the published data, reasons for the differences can only be assumed.

Differences in electrode surface treatment, handling and conditioning of insulating liquids could cause differences in measurement results between different laboratories and different researchers. It is quite demanding to define a uniform proceeding for highly reproducible measurements because influencing factors mainly depend on many individual set-up factors. For example, the waiting time prior to breakdown tests in natural ester liquids needs to be significantly higher than that in mineral oil due to its higher viscosity. Waiting times for both liquids should be long enough that a longer waiting does not affect the result significantly. Tests showed that the required waiting time strongly depends on the used oil volume. It is small for small cell volumes and high for large test tanks. Therefore, some researchers provide detailed recommendations about their experience [3]. Abrasion of electrodes strongly depends on the breakdown energy. Abrasion is significantly higher for inhomogeneous arrangements using high breakdown voltages. Suitable electrode replacement intervals have to be assured for the highest test voltage used.

Previous studies about fluid treatment of natural ester liquid showed differences to the required treatment of mineral oils that need to be respected [27].

Experiments with large volumes of insulating liquids require oil replacement management. Replacing hundreds of liters of insulating liquid after one lightning impulse breakdown needs many liters of insulating liquid which is not always required. In order to define reasonable replacement intervals, preliminary tests with at least one oil volume stressed until mean breakdown voltage starts decreasing have to be conducted.

By comparing all measured data of both setups, it can be stated that both tests were conducted considering the same findings. Differing results should be caused by the change of the conductor shape.

3.4. Comparison of Different Setups and Explanation of Their Breakdown Results

Comparisons to further test arrangements and studies investigating natural ester liquids and mineral oils at inhomogeneous field arrangements under impulse voltages are drawn to support data interpretation. Performed test study can therefore be compared to a larger database. The following studies are compared to the two analyzed setups in this contribution: [3–5,8,28,29]. First comparisons are drawn on the different electrode arrangements selected. Self-provided simplified drawings of the arrangements used are shown (if available) in Figure 9.

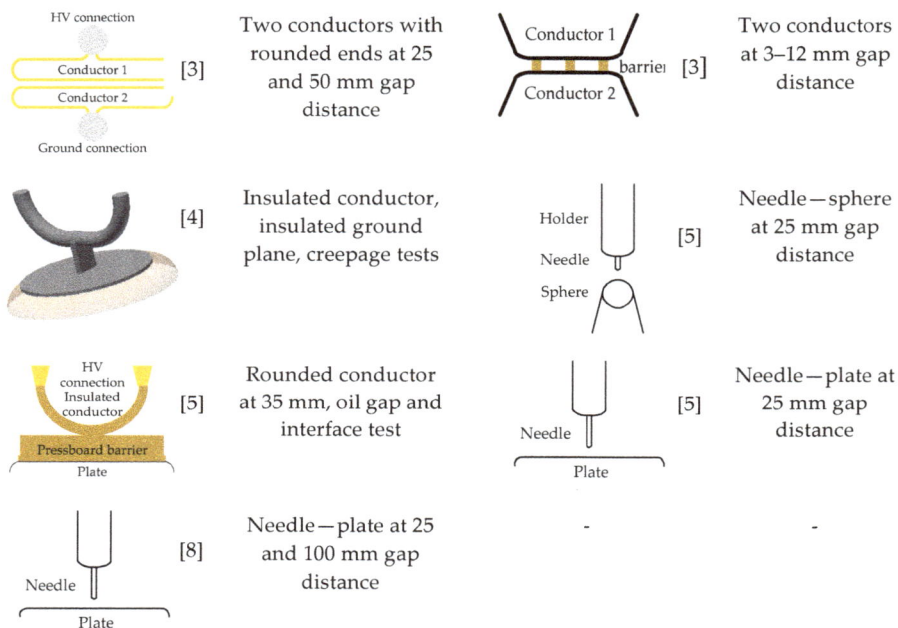

Figure 9. Different test arrangements for impulse testing of natural ester liquid compared to mineral oil.

Test arrangements can be separated into two different groups: Oil gap breakdowns with needle—plate or needle—sphere arrangements (i) and arrangements with transformer conductors (oil gap and creepage breakdowns). (ii) Breakdown voltage values and standard deviations of the compared liquids are compared for each individual group.

3.4.1. Needle Arrangement Group (i)

Group (i) consists of the arrangements of studies [5,8,29] and the present study. Liu et al. in [5] shows a comparable mean breakdown voltage for natural ester liquid and mineral oil at 25 mm oil gap under positive 1.2/50 μs impulse for needle—sphere and needle—plate arrangements. For negative impulse voltages, mineral oil shows significantly larger breakdown voltages than the natural ester liquid. Negative impulse arrangements are comparable to the present arrangements, results for needle—sphere arrangements also fit. Needle—sphere mean breakdown voltages are higher than needle—plate mean values and breakdown values under positive impulse are smaller than the ones under negative impulses (polarity effect). A comparison of breakdown voltages of the different studies is given in Figure 10.

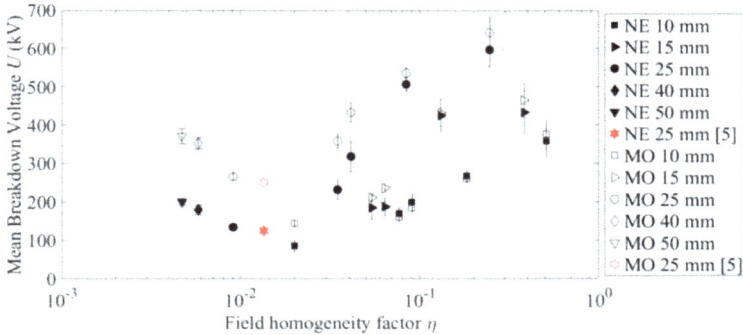

Figure 10. Comparison: mean breakdown voltages of present investigation vs. data from study [5] for natural ester liquid and mineral oil.

The electrode and gap arrangement in [8] is similar to the one presented in [4], results are also comparable. The study in [29] is also conducted with a point—plane arrangement at positive and negative impulses, but under a step voltage of 0.5/1400 μs and at 50 mm and 100 mm gap distance. Therefore, results are not directly comparable to the previous studies. They show a comparable mean breakdown voltage for natural ester liquid and mineral oil for positive step impulse at 50 mm, but a significantly reduced mean breakdown voltage of natural ester liquid (73% of the mineral oil value) at 100 mm gap distance. For negative step impulse, mean breakdown voltage of natural ester liquid is reduced significantly for both gap distances (50 mm ≈ 63% of the one of mineral oil, 100 mm ≈ 80%).

All investigated studies for group (i) have some characteristics in common: The difference in mean breakdown voltage between natural ester liquid and mineral oil at highly inhomogeneous needle configurations is small or not significant for small gaps at positive impulses. The difference between the two liquids is high for negative impulses and large gaps and can reach a 50% reduced mean breakdown voltage. The standard deviation of breakdown data of group (i) is not available from all studies. No significant difference in standard deviation for highly inhomogeneous needle arrangements could be found in the present study. Study [29] shows a significantly higher standard deviation of mineral oil, but also significantly higher breakdown values for mineral oil.

3.4.2. Conductor Arrangement Group (ii)

Group (ii) represents the arrangements of the studies [2–5] and the present study. A comparison between [2] and the present study has already been performed in Section 3.3.

U-type conductor arrangements (see Figure 9) were initially used by [4]. This arrangement can be used for oil gap or creepage breakdown investigations. T. Prevost in [4] used the arrangements for creepage investigation. Therefore, results are not directly comparable to this study because the oil-pressboard interface, is different versus the oil volume. Both polarities are investigated under 1.2/50 μs with gap distances from 10–35 mm. The results of the mean breakdown voltages for negative polarity show comparable values of natural ester liquid and mineral oil for a 10 mm gap and a slightly reduced mean value for natural ester (95% of the one of mineral oil) for 20 mm and 35 mm. For positive polarity, the largest differences between the two liquids emerge at the smallest gap distance of 10 mm with natural ester liquid showing 90% of the mean breakdown voltages of mineral oil. The differences are not significant for larger gap distances. The standard deviation of both liquids is comparable overall; some tests show a higher standard deviation for mineral oil and some tests show a higher standard deviation for natural ester liquid in the same scale.

Liu et al. in [5] used an arrangement built on the suggestion in [4] (see Figure 9) and measured oil gap breakdown and creepage breakdown at positive polarities with 1.2/50 μs impulse at 35 mm gap distance. Oil gap tests show an insignificantly reduced mean value of natural ester liquid compared to mineral oil (96%); creepage tests showed a reduced overall breakdown level of the interface systems for both liquids and a slightly reduced mean breakdown voltage for mineral oil (88%) compared to the one of natural ester liquid. Not all creepage breakdowns occurred along the interface. Differences between results of [4,5] for positive polarity impulses with creepage arrangements at 35 mm gap distance could be due to different handling of the solid-liquid insulating system in this comparatively complex setup.

Investigations in [3] composed of two opposed transformer conductors at small (3 mm) to large (50 mm) gap distance to represent a coil-to-coil arrangement instead of one conductor and a ground plate electrode as used in the previous studies. Shielded ends were applied for large gaps to prevent alternative breakdown paths. The results show a slightly reduced mean breakdown voltage of mineral oil of about 89% of the one of natural ester liquid at small gap distances (3–5 mm) and comparable results for 8–12 mm gaps. Larger gaps show an increasing difference between the two liquids with reduced mean breakdown values for natural ester liquid comparable to former mentioned studies in this range of gap distance (and respective inhomogeneities). Natural ester liquids show 88% of the mean value of mineral oil at 25 mm and 84% at 50 mm.

A general comparison between different studies using transformer conductor at 1.2/50 µs impulse voltages in comparable ranges of inhomogeneity in oil gaps shows comparable breakdown values for small gaps (condition: gap distance <25 mm). Comparison shows mainly slightly reduced breakdown voltages of natural ester liquid of around 90% of the ones of mineral oil for larger gaps (between 25 and 50 mm). There are also a few studies showing a comparable breakdown voltage for the larger gap distances. The study in this contribution shows slightly reduced mean breakdown voltages of natural ester liquid (90%) compared to mineral oil at a gap distance of 20 mm at positive polarity. The overall comparison reveals the difficulties in finding a suitable conductor arrangement built up in a simple way to avoid handling differences coming up with complex arrangements but also to reach a single wanted point of breakdown in the geometry without breakdowns at unwanted points of the geometry. The setup in [3] with shielded ends seems to be an arrangement that fulfills both requirements.

4. Conclusions

Different studies show the same trends for comparison of natural ester liquid and mineral oil under impulse voltages in inhomogeneous field arrangements. The higher the inhomogeneity, the larger the difference between natural ester liquid and mineral oil with natural ester liquid showing a reduced mean breakdown voltage. Differences in mean breakdown voltages over inhomogeneity can be explained by different streamer propagation of the two liquids built up of different chemical compounds. The larger the inhomogeneity, the higher the influence of fast streamer events leading to breakdown at lower voltages in natural ester liquids. These events occur mainly in highly inhomogeneous field arrangements made up of needle constructions ($\eta < 0.1$). Less inhomogeneous arrangements with transformer conductor ($0.1 < \eta < 0.5$) show less significant differences in fast streamer propagation and therefore less differences between the two different liquids. Typical mean breakdown values of natural ester liquid in this range of inhomogeneity are 90% of the mineral oil values. Differences are small for small gaps (e.g., gap distance 10 mm) and increase with the gap distance.

Author Contributions: Investigation, S.H.; Writing—review and editing, F.V., S.T., K.J.R. and A.S.

Funding: This research was funded by Cargill Inc., grant number EH2015-57c.

Acknowledgments: The authors would like to thank Cargill Inc. team for supporting their work.

Conflicts of Interest: The authors declare no conflict of interest. The funding sponsors had a role in the decision to publish the results, but not in the design of the study; in the collection, analyses, or interpretation of data; in the writing of the manuscript.

References

1. Pompili, M.; Calcara, L.; Sturchio, A.; Catanzaro, F. Natural esters distribution transformers: A solution for environmental and fire risk prevention. In Proceedings of the AEIT, Capri, Italy, 5–7 October 2016. [CrossRef]

2. Vukovic, D.; Jovalekic, M.; Tenbohlen, S. Comparative Experimental Study of Dielectric Strength of Oil-cellulose Insulation for Mineral and Vegetable-based Oils. In Proceedings of the IEEE ISEI, San Juan, PR, USA, 10–13 June 2012; pp. 424–428. [CrossRef]

3. Sbravati, A.; Rapp, K.J. Challenges for the application of Natural Ester fluids in extra high voltage transformers. In Proceedings of the CIGRE-AORC Technical Meeting 2016, New Delhi, India, 24–26 February 2016.

4. Prevost, T. Dielectric properties of natural esters and their influence on transformer insulation system design and performance. In Proceedings of the IEEE PES T&D, Dallas, TX, USA, 21–24 May 2006. [CrossRef]

5. Liu, R.; Törnkvist, C.; Chandramouli, V.; Girlanda, O.; Pettersson, L. Geometry impact on streamer propagation in transformer insulation liquids. In Proceedings of the Annual Report CEIDP, West Lafayette, IN, USA, 17–20 October 2010. [CrossRef]

6. Liu, Q.; Wang, Z.D. Streamer characteristic and breakdown in synthetic and natural ester transformer liquids with pressboard interface under lightning impulse voltage. *IEEE Trans. Dielectr. Electr. Insul.* **2011**, *18*, 1908–1917. [CrossRef]

7. Denat, A.; Lesaint, O.; McCluskey, F. Breakdown of liquids in long gaps: Influence of distance, impulse shape, liquid nature, and interpretation of measurements. *IEEE Trans. Dielectr. Electr. Insul.* **2015**, *22*, 2581–2591. [CrossRef]

8. Liu, R.; Törnkvist, C.; Chandramouli, V.; Girlanda, O.; Pettersson, L. Ester fluids as alternative for mineral oil: The difference in streamer velocity and LI breakdown voltage. In Proceedings of the Annual Report CEIDP, Virginia Beach, VA, USA, 18–21 October 2009; pp. 543–548. [CrossRef]

9. Hwang, J.G.; Zahn, M.; Pettersson, L. Mechanisms behind positive streamers and their distinct propagation modes in transformer oil. *IEEE Trans. Dielectr. Electr. Insul.* **2012**, *19*, 162–174. [CrossRef]

10. Lesaint, O. Prebreakdown phenomena in liquids: Propagation "modes" and basic physical properties. *J. Phys. D Appl. Phys.* **2016**. [CrossRef]

11. Lesaint, O.; Top, T.V. Streamer initiation in mineral oil. Part I: Electrode surface effect under impulse voltage. *IEEE Trans. Dielectr. Electr. Insul.* **2002**, *9*, 84–91. [CrossRef]

12. Rain, P.; Lesaint, O. Prebreakdown Phenomena in Mineral Oil under Step and Ac Voltage in Large-Gap Divergent Fields. *IEEE Trans. Dielectr. Electr. Insul.* **1994**, *1*, 692–701. [CrossRef]

13. Rozga, P.; Stanek, M. Positive Streamer Propagation in Natural Ester and Mineral Oil under Lightning Impulse. In Proceedings of the ISH, Pilsen, Czech Republic, 23–28 August 2015.

14. Rozga, P.; Stanek, M. Characteristics of streamers developing at inception voltage in small gaps of natural ester, synthetic ester and mineral oil under lightning impulse. *IET Sci. Meas. Technol.* **2016**, *10*, 50–57. [CrossRef]

15. Rozga, P.; Stanek, M. Comparative analysis of lightning breakdown voltage of natural ester liquids of different viscosities supported by light emission measurement. *IEEE Trans. Dielectr. Electr. Insul.* **2017**, *24*, 991–999. [CrossRef]

16. Lu, W.; Liu, Q. Effect of cellulose particles on impulse breakdown in ester transformer liquids in uniform electric fields. *IEEE Trans. Dielectr. Electr. Insul.* **2015**, *22*, 2554–2564. [CrossRef]

17. Weber, K.H.; Endicott, H.S. Extremal Area Effect for Large Area Electrodes for the Electric Breakdown of Transformer Oil. *Trans. Am. Inst. Electr. Eng. Part III Power Appar. Syst.* **1957**, *76*, 1091–1096. [CrossRef]

18. Haegele, S.; Tenbohlen, S.; Fritsche, R.; Rapp, K.J.; Sbravati, A. Characterization of inhomogeneous field breakdown in natural ester liquid compared to mineral oil. In Proceedings of the ICHVE, Chengdu, China, 19–22 September 2016. [CrossRef]

19. Schwaiger, A. *Elektrische Festigkeitslehre*, 2nd ed.; Springer: Berlin/Heidelberg, Germany, 1925; pp. 450–451.

20. Girdino, P.; Molfino, P.; Molinari, G. Effect of streamer shape and dimensions on local electric field conditions. *IEEE Trans. Electr. Insul.* **1988**, *23*, 4. [CrossRef]

21. Yamashita, H.; Amano, H. Prebreakdown Phenomena in Hydrocarbon Liquids. *IEEE Trans. Electr. Insul.* **1988**, *23*, 403–408. [CrossRef]

22. Forster, E.O. Critical Assessment of the Electrical Breakdown Process in Dielectric Fluids. *IEEE Trans. Electr. Insul.* **1985**, *20*, 891–896. [CrossRef]

23. Chadband, W.G.; Sufian, T.M. Experimental Support for a Model of Positive Streamer Propagation in Liquid Insulation. *IEEE Trans. Electr. Insul.* **1985**, *20*, 239–246. [CrossRef]

24. Hebner, R.E.; Kelley, E.F.; Forster, E.O.; FitzPatrick, G.J. Observations of Prebreakdown and Breakdown Phenomena in Liquid Hydrocarbons under Non-uniform Field Conditions. In Proceedings of the 1984 Eighth International Conference on Conduction and Breakdown in Dielectric Liquids, Pavia, Italy, 24–27 July 1984.

25. Lewis, T.J. Electronic Processes in Dielectric Liquids under Incipient Breakdown Stress. *IEEE Trans. Electr. Insul.* **1985**, *2*, 123–132. [CrossRef]

26. Duy, C.T.; Lesaint, O.; Denat, A.; Bonifaci, N. Streamer propagation and breakdown in natural ester at high voltage. *IEEE Trans. Dielectr. Electr. Insul.* **2009**, *16*, 1582–1594. [CrossRef]

27. Sbravati, A.; Bingenheimer, D.; Rapp, K.J.; Gomes, J.C. Fluid treatment applied for natural esters based fluids: Evaluation of fluid reconditioning and reclaiming. In Proceedings of the CIGRE Study Committee A2, Shanghai, China, 21–25 September 2015.

28. Rapp, K.J.; McShane, C.P.; Systems, C.P. Long Gap Breakdown of natural ester liquid. In Proceedings of the ICHVE 2010, New Orleans, LA, USA, 11–14 October 2010; pp. 104–107.

29. Nguyen Ngoc, M.; Lesaint, O.; Bonifaci, N.; Denat, A.; Hassanzadeh, M. A comparison of breakdown properties of natural and synthetic esters at high voltage. In Proceedings of the Annual Report CEIDP, West Lafayette, IN, USA, 17–20 October 2010. [CrossRef]

![energies logo] *energies*

MDPI

Article

Characteristics of Negative Streamer Development in Ester Liquids and Mineral Oil in a Point-To-Sphere Electrode System with a Pressboard Barrier

Pawel Rozga *, Marcin Stanek and Bartlomiej Pasternak

Institute of Electrical Engineering, Lodz University of Technology, Łódź 90-924, Poland;
marcin.stanek@dokt.p.lodz.pl (M.S.); bartlomiej.pasternak95@gmail.com (B.P.)
* Correspondence: pawel.rozga@p.lodz.pl; Tel.: +48-42-631-2676

Received: 11 April 2018; Accepted: 26 April 2018; Published: 28 April 2018

Abstract: This article presents the results of the studies on negative streamer propagation in a point-to-sphere electrode system with a pressboard barrier placed between them. The proposed electrode system gave the opportunity to assess the influence of the insulating barrier on streamer development in the conditions close to the actual transformer insulating system where the liquid gap is typically divided in parts by using pressboard barriers. The studies were performed for five commercial dielectric liquids. Among them two were biodegradable synthetic esters and two were biodegradable natural esters. Mineral oil, as the fifth liquid, was used for comparison. The measurements were based on electrical and optical experimental techniques. From the results obtained it may be concluded that, independently of the liquids tested, the electrical strength of the insulating system considered was increased by about 50%. In the case of streamer development assessed using photomultiplier-based light registration it is not possible to indicate clearly which of the liquids tested is better under the conditions of the experiment. In all cases streamers always developed slowly (2nd mode) at all voltage levels applied during the studies. In turn, the intensity of the discharge processes, comparing the same voltage levels, was mostly higher when streamers developed in ester liquids, however, the differences noticed were minimal.

Keywords: streamer propagation; mineral oil; synthetic ester; natural ester; lightning impulse; breakdown

---

1. Introduction

Ester-based insulating liquids, as eco-friendly products, have been developed providing the transformer market with an interesting alternative to mineral oils. For both natural and synthetic esters, several experimental studies have been conducted providing knowledge on different aspects of ester properties [1–7]. Without a doubt the synthetic and natural esters are friendly for the environment, being biodegradable and having high flash points. Simultaneously, they are also characterized by quite good dielectric properties, among which special attention is directed to the higher electrical permittivity of the esters than mineral oils, which is an advantage for electrical field distribution in paper-dielectric liquid insulating systems. The esters also have similar alternating current (AC) breakdown voltages to the mineral oils. In this field the benefit of the esters is their lower susceptibility to moisture content in a volume of the liquid. There is a lack of change in the AC breakdown voltage up to a moisture content of 600 ppm for synthetic esters, and up to a moisture content of 300 ppm for most of the natural esters, respectively. In turn, considering the interaction with the solid insulation in the case of ester liquids, the moisture from thermal aging of the paper is absorbed by the esters, which lengthens the paper lifetime [2,7–10].

One of the significant parameters considered in terms of dielectric properties of the liquids is the lightning impulse breakdown voltage. This parameter describes the behavior of a given liquid under lightning impulse voltage stress. This may be determined in accordance with International Electrotechnical Commission (IEC) 60897 or American Society for Testing and Materials (ASTM) D3300 standards. However, standard-based approaches limit the information only to the value of the breakdown voltage, its standard deviation and, sometimes, the time to breakdown [11–16]. Thus, with the usage of one of the above-mentioned standards, there is no possibility to assess the pre-discharge phenomena (usually called streamers) which precede the breakdown. This is possible only through the use of certain experimental techniques, among which a photomultiplier (PMT) registration of light emitted by the pre-breakdown channels is treated as one of the most valuable techniques. It is, however, a well-known fact that the intensity of light is strictly connected with the intensity of the ionization processes occurring during discharge phenomena. Developing discharges generate the pulsed light, which may be registered using proper detectors and then, in the form of waveforms, may be presented on the screen of an oscilloscope. The higher frequency of light pulses registered the more frequent step lengthening of the occurring discharge channels. The higher peak values of the light pulses reflect the higher energy of the discharge itself [15–21].

The studies which are described in this paper concern solely the mentioned PMT-based technique used for the assessment of pre-discharges (streamer development) in specific experimental electrode systems in which a point-to-sphere system recommended by the IEC 60897 standard was modified by placing an insulating plate between the electrodes. Such an approach caused the formation of a liquid gap above and below the plate. Thus, it represents, to some extent, a real situation which appears in transformer insulation—an oil gap is divided into parts by a pressboard barrier in order to limit the development of discharges which are able to form the breakdown channel. The observation of streamer development when this development is blocked by an insulating barrier became the main issue of the study.

The measurements were carried out for two synthetic esters, two natural esters, and mineral oil for comparison. Therefore, the results obtained are presented in a comparative form showing the similarities and differences between the characteristics of the liquids and the streamers developed in them under the assumed conditions of the experiment.

2. Methodology Description

The measurements were performed using the experimental setup presented schematically in Figure 1.

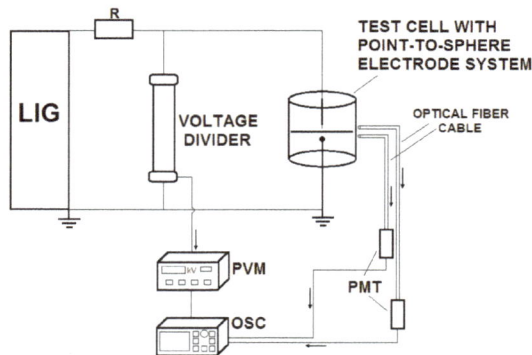

Figure 1. Laboratory setup used in the experiment: LIG—lightning impulse generator, R—limiting resistor, PMT—photomultipliers, OSC—digital oscilloscope, and PVM—peak value meter.

A source of the testing voltage was a Marx generator with a rated voltage of 500 kV and a storage energy of 2.2 kJ. It generated a negative standard lightning impulse voltage of 1.2/50 µs. The peak value of the lightning impulse was measured using a resistive voltage divider and a peak value meter. The voltage waveform was also registered by a digital oscilloscope. The voltage was supplied to the electrode system which is presented in Figure 2. This system was placed in the transparent test cell made of perspex. The volume of the test cell was 5 L.

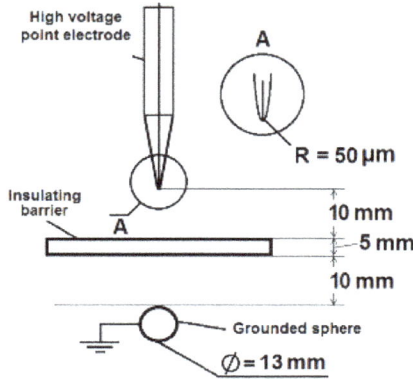

Figure 2. Electrode system used in the studies.

A high voltage needle was made of tungsten and had a 50 μm radius of curvature while the grounded electrode was a sphere 13 mm in diameter. The distance between the tip of the needle and the grounded sphere was set to 25 mm. The above-quoted dimensions of the electrodes, as well as the assumed inter-electrode distance, was based on the recommendations of the IEC 60897 standard which is used for the determination of lightning impulse breakdown voltage of dielectric liquids. The recommended electrode system was, however, modified by placing between the high voltage needle and the grounded sphere a pressboard plate 5 mm in thickness, which constituted an insulating barrier for the developed streamers. The diameter of the barrier used was 15 cm. Finally a 10 mm free liquid gap between the tip of the needle and the pressboard plate, as well as between the pressboard plate and the grounded sphere, was obtained. Before the beginning of the experiment the used pressboard plates were dried and then impregnated. The adopted procedure of drying and impregnation carried out in a vacuum chamber included:

- 24 h of drying in vacuum at a temperature of 105 °C;
- filling the chamber with a given liquid under vacuum conditions and impregnating at a temperature of 80 °C for 24 h; and
- lowering the temperature to the ambient temperature and leaving the samples for 24 h in the liquid bath.

As was mentioned above, the studies were focused on the observation of streamer development on the basis of light emission registration using the photomultiplier technique. The choice of photomultipliers as a light detector resulted from the commonly known fact that light generated by the streamers is very weak and may be detected only by very sensitive optical instruments of high gain and with a proper time response for registration of fast pulses with rise times of the order of 1–2 ns. The light from the processes observed was registered by two independently-operated photomultipliers. Both of them were Hamamatsu R1925 photomultipliers (Hamamatsu Photonics K.K., Hamamatsu, Japan) with a wavelength range from 300 to 850 nm. The light signal was delivered to the PMTs through optical fiber cables which consisted of 37 individual fibers. The active surface of the cables collecting the light from the processes registered was 5 mm in diameter. The ends of both cables were placed directly at the wall of the test cell at a distance from the point-sphere axis of ca. 10 cm. As it was presented in Figure 1, the end of one of the optical fiber cables was placed above the line defined by the surface of the insulating plate while the end of the second optical fiber cable was placed below the quoted line. In other words, the first PMT recorded light from the processes occurring between the high-voltage needle and the insulating barrier, while the second PMT recorded the processes occurring between the insulating barrier and the grounded sphere. The output signals from both the PMTs, after being amplified, were registered using the oscilloscope in the form of waveforms (the temporal sequence of discrete negative pulses having rise-times of nanoseconds). Simultaneously, the waveforms obtained were saved on a hard disk of a PC computer where they might be analyzed and compared with the other measurement results. During the whole experiment the PMTs were supplied by the same supply voltage (800 V), which means that the light was acquired with the same gain factor for both the PMTs. Additionally, the same types of amplifiers were used to amplify the output signal from the photomultipliers.

Five different commercial dielectric liquids were tested during the experiment. They included two synthetic esters, two natural esters, and mineral oil for comparison. The data concerning the fundamental physico-chemical and dielectric properties of the liquids tested are presented in Table 1. These data come from the datasheets provided by the manufacturers or from other available documents [1,2,22–25].

Table 1. Basic dielectric parameters of the liquids tested.

Parameters	Synthetic Ester I	Synthetic Ester II	Natural Ester I	Natural Ester II	Mineral Oil
Density at 20 °C (kg/dm$^{3)}$	0.97	0.97	0.92	0.89	0.88
Viscosity at 40 °C (mm^2/s)	29	29	32	17	10
Viscosity at 100 °C (mm^2/s)	5.25	5.6	7.7	4.6	2.6
Flash point (°C)	260	265	320	200	191
Pour point (°C)	−56	−54	−23	−28	−42
Biodegradability	readily biodegradable	readily biodegradable	readily biodegradable	readily biodegradable	not biodegradable
Moisture saturation in ambient temperature (ppm)	2700	2700	1100	no data	55
AC breakdown voltage—mean value (kV)	>75	78	73	70	70
Dielectric dissipation factor at 90 °C and 50 Hz	<0.02	<0.02	<0.02	0.04	<0.002
Electrical permittivity at 20 °C	3.2	3.2	3.1	2.8	2.2

As can be seen in Table 1, the synthetic esters tested have similar properties in both the chemical and dielectric range, and are of similar origin with respect to the manufacturing process. The natural esters tested have the same vegetable origin, but are characterized by different values of some fundamental parameters. Firstly, natural ester II has lower viscosity, which is not a typical property for natural esters for electrical purposes. Like other natural esters it is based on triglycerides, but their content is reduced in the direction of using more monoesters. This is made by chemical modifications to the liquid. The ratio between the content of triglycerides and monoesters (in the case of the considered ester it is ca. 50% to 50%) which drives the balance between the properties of the final product. With this, the viscosity was optimized by reducing the flash point, but still keeping the biodegradability. Finally the reference liquid-mineral oil-is of a naphthenic type and is uninhibited.

Prior to the measurements carried out at the lightning impulse voltage, the mineral oil tested was first preprocessed by drying, degassing, and filtering. However, the ester liquids tested were studied as received. In order to be sure that the liquids fulfill the criteria specified individually for a given liquid the quality of the liquids was verified by AC breakdown voltage tests using the IEC 60156 standard. For all the liquids considered, the average values of the AC breakdown voltage exceed 60 kV, which confirmed the appropriate properties of the liquids for lightning tests. In turn, the moisture content of the liquid samples was also measured using IEC 60814. The values obtained were, respectively, 12 ppm for mineral oil, 128 ppm for synthetic ester I, 88 ppm for synthetic ester II, 62 ppm for natural ester I, and 49 ppm for natural ester II. According to the well-known fact that higher natural moisture saturation of the esters and the above-mentioned lack of influence of moisture content on the breakdown voltage of the esters, the data concerning moisture were also assessed as satisfactory.

The measurement procedure was established on the basis of previously-performed studies on lightning impulse breakdown voltage determination [15,16]. Schematically, the used procedure is presented in Figure 3.

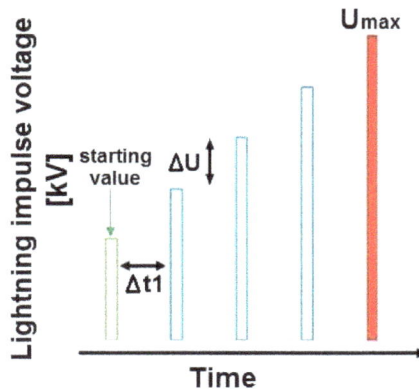

Figure 3. Schematic representation of the measurement procedure; $\Delta t1$—time between distinctive voltage steps, ΔU—assumed voltage step, and U_{max}—assumed maximum value of the testing voltage.

The experiment began from the so-called starting value, which was 100 kV. This value was lower than previously-evaluated negative lightning impulse breakdown voltages for all the liquids tested in point-to-sphere electrode systems [15,16]. Since the evaluated negative lightning impulse breakdown voltage oscillated around 125 kV with ca. 5–7 kV of standard deviation, it was assumed that a value of 100 kV would be the right starting value at which streamers will develop as a basic, slow mode [18–21]. Then the subsequent lightning impulses were supplied to the electrode system with the assumed voltage step (ΔU) equal to 5 kV. One shot per step was applied with time between the subsequent steps ($\Delta t1$) equal to 1 min. The measurements were stopped at the value of the testing voltage equal to 180 kV (U_{max} in Figure 3). The next measurement procedure was carried out after a 20 min break wherein during this time the liquid sample was stirred. Based on our own experience [15,16,19], as well as recommendations of the standards [11,12], it was decided that the liquid samples would not be changed to a new one after a given measurement procedure. This was because breakdown was not reached during the tests. The confirmation of the correctness of the approach adopted was an unintended spread of the data obtained from the measurements. The influence of the previous test on the results obtained from the next measurement procedure was not noticed. The total number of measurement procedures was 6. For each voltage level the oscillograms were registered. They included two courses of light (separately for both the PMTs) and one course of supplied voltage. In this latter case it was possible to identify probable breakdown events [15,16,18,21,26]. However, for all the voltage levels applied during the experiment the voltage collapse in the voltage waveforms registered never took place, so it may be stated that breakdown never happened.

3. Measurement Results

Firstly, the analysis was focused on the comparison of the threshold value of the voltage at which the light signal was registered by the PMT that was placed below the insulating barrier. This value was called V_t, meaning that the streamers also started to propagate in the space between the insulating barrier and grounded sphere. As was mentioned above, six measurement procedures were performed so the V_t in each case was determined as an average value calculated from the six individual observations. Simultaneously, standard deviations were also evaluated. The results obtained are shown in Table 2. In order to relate these results to the previously-estimated negative lightning impulse breakdown voltages (V_{b-}) for a point-to-sphere electrode system without a pressboard barrier [15,16], the table also includes the mentioned V_{b-}.

Table 2. Characteristic voltages for the discharge phenomena in a point-to-sphere electrode system with and without an insulating barrier.

	Parameters				
	V_t (kV) for Point-To-Sphere Electrode System with Insulating Barrier		V_{b-} (kV) for Point-To-Sphere Electrode System without Insulating Barrier		V_t/V_{b-} Ratio
Type of the Liquid	Average Value	Standard Deviation	Average Value	Standard Deviation	-
Synthetic ester I	135.83	5.84	123.00	5.23	1.10
Synthetic ester II	140.83	9.17	121.50	5.10	1.16
Natural ester I	132.50	10.37	123.25	4.94	1.07
Natural ester II	127.50	8.21	120.75	5.91	1.05
Mineral oil	130.00	9.49	127.50	6.98	1.02

The results from the table are also presented in graphical form in Figure 4.

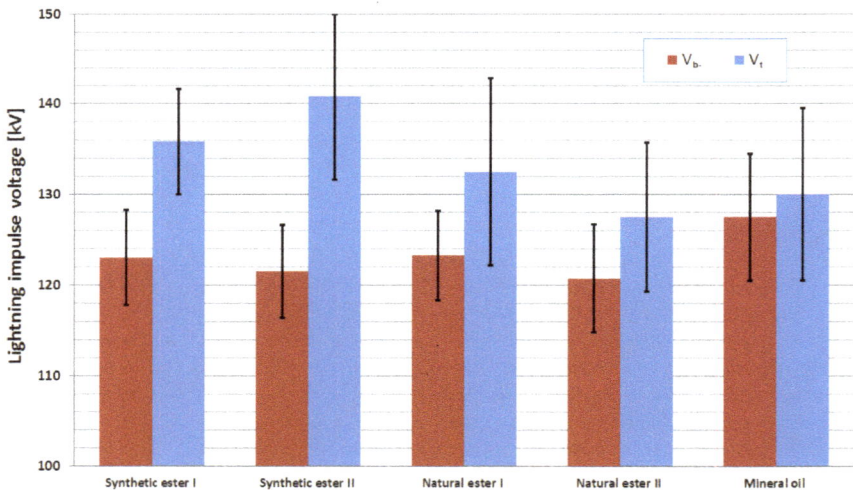

Figure 4. Graphical representation of the measurement results: V_{b-}: negative lightning impulse breakdown voltage from previous studies, and V_t: threshold value of voltage at which the light signal was registered by the PMT placed below the insulating barrier.

Comparing the results concerning the threshold value of voltage for streamer inception below the insulating barrier with the values of the lightning impulse breakdown voltage obtained during the previous studies, it may be said that, in all cases, the values of V_t are higher than the corresponding values of V_{b-} (the V_t/V_{b-} ratio quoted in Table 2 is, in all cases, much higher than 1). Hence, the conclusion in this field is not surprising—the insulating barrier prevents the breakdown which would certainly take place at voltage levels equal to V_t, or close to this value. This applies to the same extent to all the liquids tested. Nevertheless, comparing the distinctive liquids between each other there are obvious differences in the values of evaluated V_t, which are confirmed by the individual V_t/V_{b-} ratios. The best properties mean the best interaction of a given liquid with the pressboard plate in terms of lightning strength is shown by synthetic ester II. The next, in turn, are synthetic ester I and natural ester I. In the case of mineral oil and natural ester II, the properties are similar. Comparing the real value of V_t, mineral oil has better properties, but looking at the V_t/V_{b-} ratio, natural ester II seems to be better. Simply, it may be concluded that the liquids of higher electrical permittivity (both synthetic esters tested) behave better under the conditions of the experiment. Simultaneously, it is important to point out that, in any of the liquids tested, breakdown was not noticed. Thus, the improvement of the electrical strength of the insulating system analyzed took place in a similar range—the lightning breakdown voltage of the system increased from around 125 kV, up to 180 kV. It may be supposed that the reason for this is the more favorable electrical field distribution in the synthetic ester-pressboard insulating system than in the case of mineral oil-pressboard systems. Higher electrical permittivity of the synthetic esters, which is closer to the electrical permittivity of the pressboard, causes lower stresses on the border of the pressboard and liquid. Hence, the lower the electrical stress is, the higher the value of the voltage that is needed to initiate streamers below the pressboard barrier.

The next part of the study concerns the analysis of the light oscillograms. As was mentioned above, the oscillograms were registered at each voltage level. In general, for each of the liquids tested, the sequence was such that, for the first few levels, the voltage light signal (discrete pulses increasing in time) was registered only by the PMT placed above the insulating plate. After reaching a given voltage level (meaning V_t) the pulses started to also appear in the course connected with the second PMT (channel No. 2). In the case of both channels, with an increase of the voltage level, the frequency of light pulses also increased. The sequence of the oscillograms registered with the voltage increase is shown in Figure 5 and the example chosen is the synthetic ester II.

The appearance of the streamers on the lower side of the insulating barrier results from the conditions which occurred in the space between barrier and grounded sphere after exceeding a given value of the testing voltage. These conditions were forced by electrical field stress in the mentioned space, which enabled the streamer initiation. At lower values of the testing voltage it was not possible to initiate the streamers below the insulating barrier because the distance between this area and the high-voltage point electrode was too long and the geometric electrical field resulting from the electrode configuration and the applied voltage was too low. In such conditions even the space charge from the discharge developed above the insulating barrier did not disturb the field strongly enough to cause streamer initiation in the area close to the sphere electrode. For the higher values of the testing voltages, when the light pulses are visible in course No. 2 connected with the PMT 2 placed below the insulating barrier, the geometric electrical field stress existing below the insulating plate is high enough, and by deformation of this field distribution by the appearance of the space charge it becomes sufficient for ionization processes, initiating streamer development (because of the high value of the local electrical field the streamers started to develop at the weak points occurring in the liquid volume). Thus, for lower testing voltages, although the streamers developed above the barrier, none of the phenomena were observed below it. Only by increasing the voltage, which causes a proportional increase in the value of the electric field stress in the entire space between the point high voltage (HV) electrode and the grounded sphere, the discharge propagation was noticed below the pressboard barrier. The next voltage levels strengthen the phenomena, which was confirmed by the more intense light registered by the PMTs. The above-presented considerations concern, to the same extent, all the liquids tested.

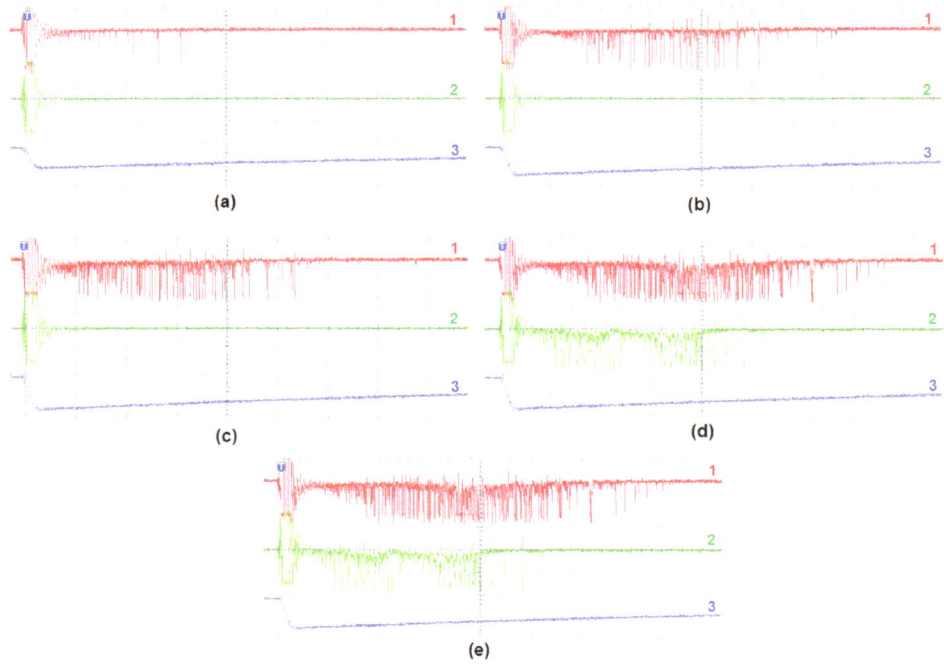

Figure 5. The sequence of streamer development with the voltage increase for synthetic ester II: (**a**) *V* = 100 kV; (**b**) *V* = 125 kV; (**c**) *V* = 140 kV; (**d**) *V* = 145 kV; and (**e**) *V* = 170 kV; 1—light from PMT 1 (arb. units), 2—light from PMT 2 (arb. units), 3—voltage (100 kV/div.), and *t* = 4 μs/div.

What is additionally interesting in the described situation is that the signal registered by the PMT 2 includes some kind of continuous state visible, for example, in Figure 5d,e. An explanation of this fact may be the effect of the space charge which accumulated on the surface or inside the pressboard plate. When the electrical field affects the space around the plate and, of course, the plate itself, it is possible to excite this space charge for ionization processes, which is reflected in the emitted, and then registered, light.

What is also important to point out is that, in all cases and for all ranges of the testing voltage applied, the streamers always propagated as the slow-type of discharge. The time of streamer development was relatively long and the phenomena characteristics for the fast-type of discharges as in [17–21,26] were not noticed.

The comparative assessment of the liquids tested may be done when we consider the set of the oscillograms for distinctive liquids, which were registered at the same voltage levels. Such a comparison is made in Figures 6 and 7, respectively, for two cases: when streamers developed only above the insulating barrier and when the streamers were observed on both sides.

In the first case, that is, for the oscillograms presented in Figure 6, the main observations are identical to those noticed during the studies in the point-to-sphere electrode system without an insulating barrier [15,16]. The difference between the esters and the mineral oil is that in mineral oils the light pulses are characterized by lower peak values and a slightly lower frequency. More detailed analysis allowed formulating the statement that, in the case of both synthetic esters, the discharges propagate longer than in the case of the natural esters tested and the mineral oil. When we set together the oscillograms registered at the same value of the testing voltage at, for example, 125 kV, as presented in Figure 6, the average time of discharge development for both synthetic esters is ca. 20 μs, while the same time corresponding with natural esters and mineral oil is 12–14 μs. On this basis it may be concluded that the conditions for streamer propagation (lengthening of the streamer channels) under the conditions of the experiment, are slightly better in the case of synthetic esters.

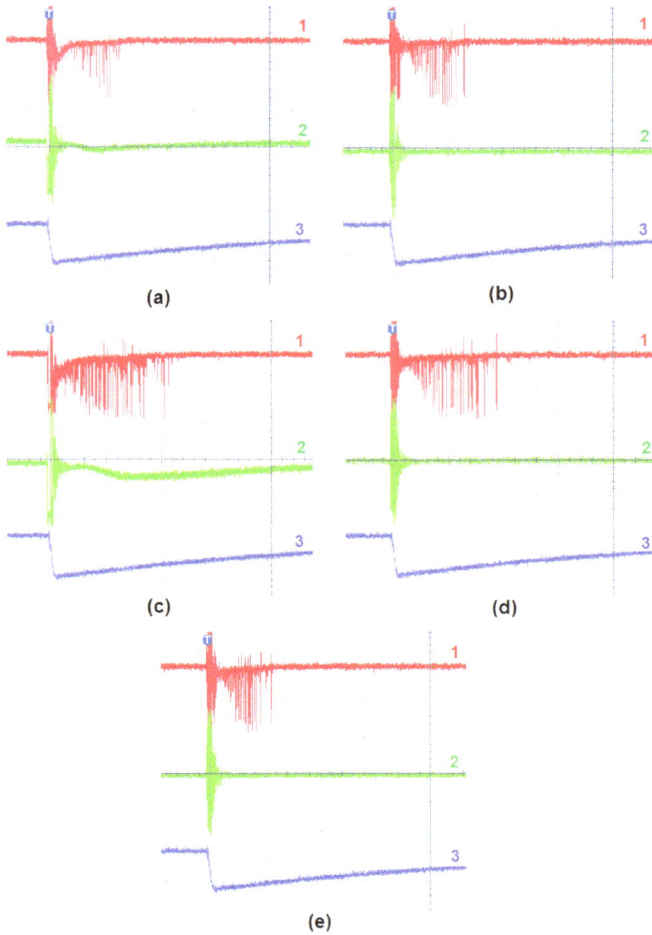

Figure 6. The set of oscillograms registered in the case of streamer development only above the insulating barrier, $V = 120$ kV: (**a**) mineral oil; (**b**) natural ester I; (**c**) synthetic ester I; (**d**) synthetic ester II; and (**e**) natural ester II; 1—light from PMT 1 (arb. units), 2—light from PMT 2 (arb. units), 3—voltage (100 kV/div.), and $t = 10$ μs/div.

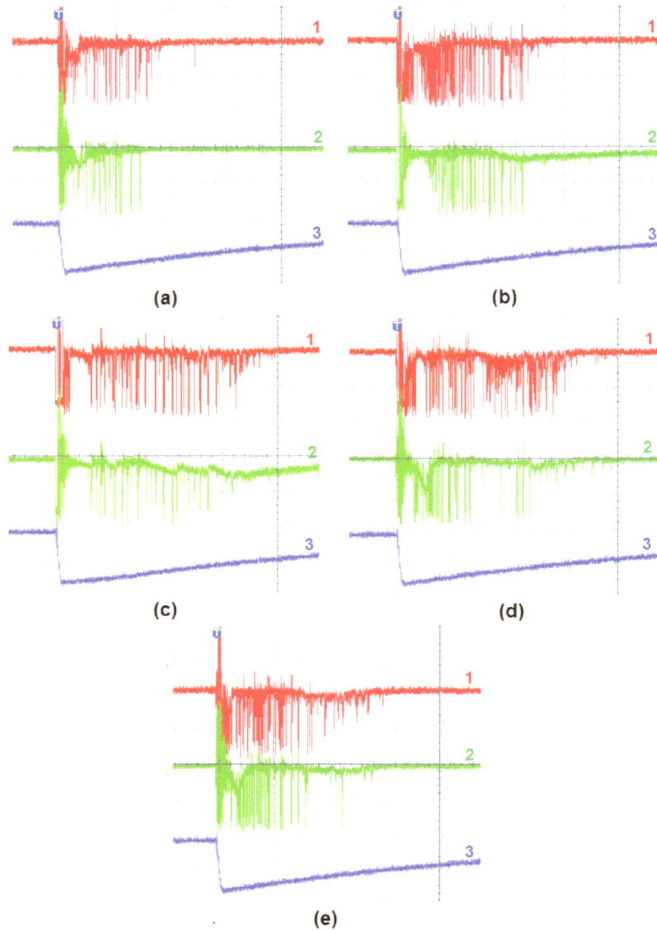

Figure 7. The set of oscillograms registered in the case of streamer development below and above the insulating barrier, V = 150 kV: (**a**) mineral oil; (**b**) natural ester I; (**c**) synthetic ester I; (**d**) synthetic ester II, and (**e**) natural ester II; 1—light from PMT 1 (arb. units), 2—light from PMT 2 (arb. units), 3—voltage (100 kV/div.), and t = 10 µs/div.

In turn, comparing with each other the oscillograms from Figure 7, the conclusions are, in a sense, consistent with the observations based on Figure 6. The time of discharge development in mineral oil is the shortest. A slightly longer time is noticed for both natural esters, while the longest time concerns the synthetic esters tested. Assessing, however, the waveforms registered by the PMT placed above and below the insulating barrier, it may be noticed that there is some correlation between the intensity of the light pulses registered for both sides. Thus, there is also a correlation between the processes occurring above and below the barrier after exceeding a given threshold value of the testing voltage. The typical waveforms are characterized by similar frequencies and peak values independently whether the streamers develop above or below the pressboard plate. This is, in fact, similarly visible for all the liquids tested. Hence, it may be said that the electrical field stress above the barrier (mainly resulting from the applied voltage and the radius of curvature of the HV point electrode) is very close in value to the electrical field stress below the barrier, which is disturbed by the existing space charge. However this conclusion is not unequivocal, because in some cases the phenomena occurring below the barrier, represented by the light pulses, are less intense. The pulses are less frequent and have lower peak values.

In conclusion, independently of the liquid type, it may be stated that insulating pressboard barriers effectively prevent the development of the discharges leading to breakdown. Probably, the discharges propagate in both areas in the direction of the insulating plate and accumulate on its surface. Due to the high electrical strength of the pressboard, the intensity of the discharges in all cases are too low to cause the formation of a breakdown channel. At least up to a peak value of 180 kV of the standard lightning impulse voltage of negative polarity, such breakdown is not possible. Thus, all the liquids tested, although the existing differences in the lightning impulse breakdown voltages determined using the IEC 60897 standard, as well as the differences in the threshold value of the voltage at which the streamers started to develop below the insulating barrier (V_t), behaved similarly under the conditions of the experiment, allowing for an efficient stopping of the formation of the discharge of a breakdown nature.

4. Conclusions

On the basis of the studies performed, the following conclusions can be drawn:

(1) Independently of the liquid tested, the insulating barrier similarly improves the electrical strength of the tested electrode system. The barrier prevents the breakdown up to 180 kV (the maximum applied voltage), which is almost 50% higher than the breakdown voltage measured in a point-to-sphere electrode system.

(2) Comparing the obtained V_t/V_{b-} ratios the synthetic esters tested seem to improve the lightning properties of the tested electrode system to the greatest extent. Since a longer time of discharge development was registered in the case of both synthetic esters at all testing voltage levels, it may be stated that the conditions for propagation of the discharges in synthetic esters are, however, slightly easier than for other tested dielectric liquids (mineral oil and natural esters). Thus, it is not possible to indicate clearly which of the liquids tested is better under the conditions of the experiment.

(3) From the registration of light using two photomultipliers it may be said that the relationships between the liquids are similar to the case of the system without an insulating barrier:

- streamers develop slowly (2nd mode) at all voltage levels applied during the studies; and
- the intensity of discharge processes, comparing the same voltage levels, was always higher when streamers developed in ester liquids.

(3) Generally, the results obtained showed that, in real insulting systems of power transformers where the oil gap is divided into parts using pressboard barriers, the use of ester liquids as an insulating medium does not necessarily generate the problems of discharge nature when considering lightning impulse stresses.

Author Contributions: Pawel Rozga was the originator of the studies and their coordinator. He also participated in the interpretation of the results, forming the conclusions, and in writing the paper. Marcin Stanek and Bartlomiej Pasternak performed the studies and analyzed their results. Pawel Rozga and Marcin Stanek wrote the paper.

Conflicts of Interest: The authors declare no conflict of interest.

References

1. Gockenbach, E.; Borsi, H. Natural and synthetic ester liquids as alternative to mineral oil for power transformers. In Proceedings of the 2008 Annual Report Conference on Electrical Insulation and Dielectric Phenomena (CEIDP 2008), Quebec, QC, Canada, 26–29 October 2008; pp. 521–524. [CrossRef]

2. Perrier, C.; Beroual, A. Experimental investigations on insulating liquids for power transformers: Mineral, ester and silicone oils. *IEEE Electr. Insul. Mag.* **2009**, *25*, 6–13. [CrossRef]

3. Talhi, M.; Fofana, I.; Flazi, S. Comparative study of the electrostatic charging tendency between synthetic ester and mineral oil. *IEEE Trans. Dielectr. Electr. Insul.* **2013**, *20*, 1598–1606. [CrossRef]

4. Fofana, I. 50 years in the development of insulating liquids. *IEEE Electr. Insul. Mag.* **2013**, *29*, 13–25. [CrossRef]

5. Bandara, K.; Ekanayake, C.; Saha, T.; Ma, H. Performance of natural ester as a transformer oil in moisture-rich environments. *Energies* **2016**, *9*, 258. [CrossRef]

6. Beroual, A.; Khaled, U.; Mbolo Noah, P.S.; Sitorus, H. Comparative study of breakdown voltage of mineral, synthetic and natural oils and based mineral oil mixtures under AC and DC Voltages. *Energies* **2017**, *10*, 511. [CrossRef]

7. Fernández, O.H.A.; Fofana, I.; Jalbert, J.; Gagnon, S.; Rodriguez-Celis, E.; Duchesne, S.; Ryadi, M. Aging characterization of electrical insulation papers impregnated with synthetic ester and mineral oil: Correlations between mechanical properties, depolymerization and some chemical markers. *IEEE Trans. Dielectr. Electr. Insul.* **2018**, *25*, 217–227. [CrossRef]

8. Liao, R.; Hao, J.; Chen, G.; Ma, Z.; Yang, L. A comparative study of physicochemical, dielectric and thermal properties of pressboard insulation impregnated with natural ester and mineral oil. *IEEE Trans. Dielectr. Electr. Insul.* **2011**, *18*, 1626–1637. [CrossRef]

9. Lashbrook, M.; Kuhn, M. The use of ester transformer fluids for increased fire safety and reduced costs. *CIGRE Tech. Progr.* **2012**, A2–A210.

10. Martins, M.A.G.; Gomes, A.R. Comparative study of the thermal degradation of synthetic and natural esters and mineral oil: effect of oil type in the thermal degradation of insulating Kraft paper. *IEEE Electr. Insul. Mag.* **2012**, *28*, 22–28. [CrossRef]

11. IEC 60897. *Methods for the Determination of the Lightning Impulse Breakdown Voltage of Insulating Liquids*; IEC: Geneva, Switzerland, 1987.

12. ASTM D3300-12. *Standard Test Method for Dielectric Breakdown Voltage of Insulating Oils of Petroleum Origin under Impulse Conditions*; ASTM International: West Conshohocken, PA, USA, 2012.

13. Liu, Q.; Wang, Z.D.; Perrot, F. Impulse breakdown voltages of ester-based transformer oils determined by using different test methods. In Proceedings of the 2009 IEEE Conference on Electrical Insulation and Dielectric Phenomena (CEIDP '09), Virginia Beach, VA, USA, 18–21 October 2009; pp. 608–612. [CrossRef]

14. Peppas, G.D.; Charalampakos, V.P.; Pyrgioti, E.C.; Gonos, I.F. Electrical and optical measurements investigation of the pre-breakdown processes in natural ester oil under different impulse voltage waveforms. *IET Sci. Meas. Technol.* **2016**, *10*, 545–551. [CrossRef]

15. Rozga, P.; Stanek, M. Comparative analysis of lightning breakdown voltage of natural ester liquids of different viscosities supported by light emission measurement. *IEEE Trans. Dielectr. Electr. Insul.* **2017**, *24*, 991–999. [CrossRef]

16. Stanek, M.; Rozga, P.; Rapp, K. Comparison of lightning characteristics of selected insulating synthetic esters with mineral oil. In Proceedings of the 2017 IEEE 19th International Conference on Dielectric Liquids (ICDL), Manchester, UK, 25–29 June 2017; pp. 1–4. [CrossRef]

17. Beroual, A.; Zahn, M.; Badent, A.; Kist, K.; Schwabe, A.J.; Yamashita, H.; Yamazawa, K.; Danikas, M.; Chadband, W.D.; Torshin, Y. Propagation and structure of streamers in liquid dielectrics. *IEEE Electr. Insul. Mag.* **1998**, *14*, 6–17. [CrossRef]

18. Liu, Q.; Wang, Z.D. Streamer characteristic and breakdown in synthetic and natural ester transformer liquids under standard lightning impulse voltage. *IEEE Trans. Dielectr. Electr. Insul.* **2011**, *18*, 285–294. [CrossRef]

19. Rozga, P. Streamer propagation and breakdown in very small point-insulating plate gap in mineral oil and ester liquids at positive lightning impulse voltage. *Energies* **2016**, *9*, 467. [CrossRef]

20. Rozga, P. Using the light emission measurement in assessment of electrical discharge development in different liquid dielectrics under lightning impulse voltage. *Electr. Power Syst. Res.* **2016**, *140*, 321–328. [CrossRef]

21. Lesaint, O. Prebreakdown phenomena in liquids: propagation modes and basic physical properties. *J. Phys. D Appl. Phys.* **2016**, *49*, 14401–14422. [CrossRef]

22. Szewczyk, R.; Vercesi, G. *Innovative Insulation Materials for Liquid-Immersed Transformers*; DuPont Webinar: Virginia Beach, VA, USA, 13 March 2015.

23. Envirotemp FR3 Fluid Brochure. Available online: https://www.cargill.com/bioindustrial/dielectric-fluids (accessed on 6 April 2018).
24. Envirotemp 200 Fluid Brochure. Available online: https://www.cargill.com/bioindustrial/dielectric-fluids (accessed on 6 April 2018).
25. Midel 7131 Product Brochure. Available online: https://www.midel.com/ (accessed on 6 April 2018).
26. Duy, C.T.; Denat, A.; Lesaint, O.; Bonifaci, N. Streamer propagation and breakdown in natural ester at high voltage. *IEEE Trans. Dielectr. Electr. Insul.* **2009**, *16*, 1582–1594. [CrossRef]

energies

MDPI

Article

Electronic Properties of Typical Molecules and the Discharge Mechanism of Vegetable and Mineral Insulating Oils

Yachao Wang, Feipeng Wang *, Jian Li *, Suning Liang and Jinghan Zhou

State Key Laboratory of Power Transmission Equipment & System Security and New Technology,
School of Electrical Engineering, Chongqing University, Chongqing 400044, China;
wangyachao@cqu.edu.cn (Y.W.); 20141101004@cqu.edu.cn (S.L.); lxr1236@126.com (J.Z.)
* Correspondence: fpwang@cqu.edu.cn (F.W.); lijian@cqu.edu.cn (J.L.);
Tel.: +86-023-65102434 (F.W.); +86-023-6510-2437 (J.L.)

Received: 25 January 2018; Accepted: 24 February 2018; Published: 28 February 2018

Abstract: Vegetable insulating oil may replace the mineral insulating oil used in large power transformers due to its extraordinary biodegradability and fire resistance. According to component analysis, 1-methylnaphthalene and eicosane are considered the typical molecules in mineral oil. Triolein and tristearin are considered the typical molecules in vegetable oil. The ionization potential (IP) and the variation of highest occupied molecular orbital (HOMO) of typical molecules under an external electric field are calculated using quantum chemistry methods. The calculation results show that the IP of the triolein molecule is comparable to that of the 1-methylnaphthalene molecule. The mechanisms of losing electrons are discussed, based on the analysis of HOMO composition. The insulation characteristics of the triolein and tristearin are more likely to be degraded under an external electric field than those of 1-methylnaphthalene and eicosane. Due to the fact that the number density of low IP molecules groups in vegetable oil is much greater than that in mineral oil, the polarity effect in vegetable oil is more obvious than that in mineral oil. This eventually leads to different streamer characteristics in vegetable oil and mineral oil under positive polarity and negative polarity.

Keywords: vegetable oil; mineral oil; electronic property; streamer; space charge; density functional theory

1. Introduction

Vegetable insulating oil is a nontoxic, reproducible and environmental dielectric fluid [1–7]. It can be used as a substitute for mineral insulating oil, in transformer and oil-filled cables [8–11]. The basic physical, chemical and electrical properties of vegetable oil and mineral oil are listed in Table 1 [3]. It is attracting increasing research interest due to its high fire point of above 300 °C [4], which significantly improves the safety level of the power grid. Moreover, the 21-day degradation rate of vegetable oil is up to 97% by the CEC-L-33 test (Coordinating European Council, Leicester), which is superior to mineral oil with 30% [5]. Generally speaking, insulating materials are very sensitive to moisture, such as insulation paper and low-density polyethylene [12,13]. The frequency breakdown voltage of mineral oil will collapse when the moisture is up to 30 ppm, however, the frequency breakdown voltage of vegetable oil will not obviously decrease when the moisture is up to 320 ppm [6,14]. Vegetable oil absorbs the moisture in insulation paper, leading to much more prolonged lifetime of oil-paper insulations [7].

Table 1. Basic physical, chemical and electrical properties of vegetable oils and a mineral oil [3].

Parameter	Camellia Oil	FR3 Oil	Mineral Oil
Appearance	Light Yellow	Light Green	Transparent
Density (20 °C)/kg·m^{-3}	0.90	0.92	<0.895
Viscosity (40 °C)/mm^2·s^{-1}	39.9	34.1	≤13.0
Pour point/°C	−28	−21	<−22
Flash point/°C	322	316	≥135
Acid value/mgKOH·g^{-1}	0.03	0.04	≤0.03
Interfacial tension/mN·m	25	24	≥40
AC breakdown voltage/kV	70	56	≥35
Dissipation factor(90 °C)/%	0.88	0.89	≤0.1
Volume resistivity/Ω·m	1×10^{10}/90 °C	2×10^{11}/25 °C	7×10^{11}/25 °C
Relative permittivity	2.9/90 °C	3.2/25 °C	2.2/90 °C

Vegetable oil has shown comparable electrical properties to mineral oil by some standard tests [8]. However, comparing with mineral oil, fast streamer seems to appear easily in vegetable oil especially when the oil gap is longer than 50 mm. Liu's study shows that [15], under positive polarity, the streamer velocity in vegetable oil is about 10 km/s at 50 mm of oil gap and reaches 30 km/s at 100 mm. The streamer velocity in mineral oil remains at a constant of 1–2 km/s for oil gaps of 25–100 mm under positive polarity. However, under negative polarity, the streamer velocity in vegetable oil seems to be suppressed and the streamer velocity in mineral oil obviously increases with the increasing oil gap. For example, the streamer velocity in mineral oil is about 6 km/s at 75 mm of oil gap under negative polarity, even higher than that in vegetable oil. The fast streamer (>10 km/s) in the insulating liquid may connect the electrodes, resulting in serious disaster. However, the streamer propagation is affected by many factors such as air-pressure, temperature, impurities and moisture. The mechanisms of streamer propagation are not completely understood.

To date, considerable results have focused on the assessment of differences and similarities between vegetable oil and mineral oil based on experimental measurements and analysis. There is a lack of investigation at the molecular level into explaining the electrical discharge phenomena happening in the oils. Density functional theory (DFT) is a quantum chemistry method to calculate the electronic structure. DFT is widely used to speculate about the properties of molecular and condensed matter in both physics and chemistry [16,17]. The molecular components of insulating oils are very plentiful and we chose several typical molecules to study the molecular characteristics associated with electric discharge.

Mineral oil, as transformer oil, is from petroleum by distillation at a temperature of 260–400 °C. According to the component analysis, mineral oil is dominated by hydrocarbons that include the aromatic, paraffinic and naphthenic [18–20]. Aromatic usually has two benzene rings (e.g., 1-methylnaphthalene molecule) and the total proportion of aromatic is about 5%. Paraffinic and naphthenic usually have 16~22 carbon units (e.g., eicosane molecule) and the total proportion of paraffinic and naphthenic is about 95%. For vegetable oil, the component analysis has indicated that the main molecules are triglycerides [21–25]. The triglyceride molecule can be considered a glycerol molecule esterified by three fatty acid molecules. The fatty acids usually have 14~22 carbon units. The unsaturated fatty acids (e.g., oleic acid, generating triolein molecule in vegetable oil) hold the majority among the fatty acids (about 90%). A small amount of saturated fatty acids (e.g., stearic acid, generating tristearin molecule in vegetable oil) are included as well (about 10%).

In this work, according to the component analysis, 1-methylnaphthalene and eicosane are considered the typical molecules in mineral oil. Triolein and tristearin are considered the typical molecules in vegetable oil. The electronic properties of typical molecules are calculated using the quantum chemistry method. The mechanisms of fast and slow streamers in vegetable oil and mineral oil are discussed based on the calculations.

2. Methods

According to the component analysis, the four typical molecules are considered including 1-methylnaphthalene, eicosane, triolein and tristearin. The models for the typical molecules with a Cartesian axis are shown in Figure 1.

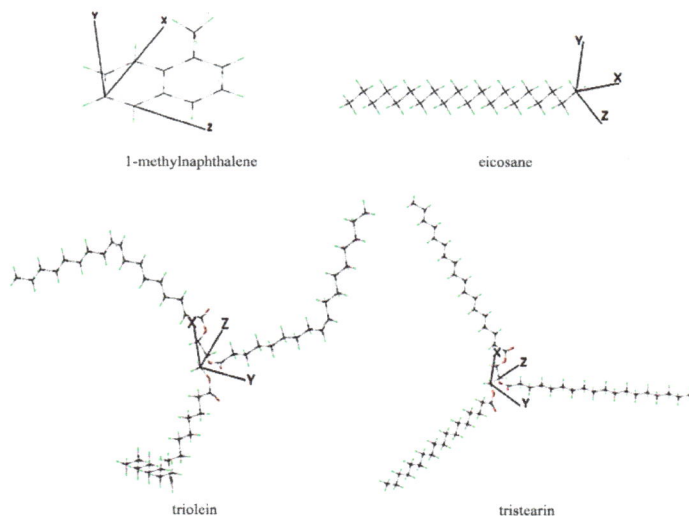

1-methylnaphthalene   eicosane

triolein   tristearin

Figure 1. Models of typical molecules with Cartesian axis. Gray: C, red: O, green: H.

The IPs of the typical molecules are calculated by quantum chemistry methods based on density functional theory. The calculations are carried out using the Gaussian 09 program package [26]. The B3LYP method is employed at 6-31G* basis set [27,28].

The geometries of the molecules are optimized for energy minimization. The energy of a positive ion is calculated based on the geometry optimized. The *IP* of molecule *A* is defined as the energy difference of neutral molecule *A* and corresponding positive ion A^+, written as:

$$IP = E^{A^+} - E^A \tag{1}$$

The polarizable continuum model (PCM) is used to calculate the *IP* [29]. In PCM, the surrounding liquid for a given molecule is considered as a continuum dielectric with a dielectric constant ε. It is set to be 3.0 for vegetable oil and 2.2 for mineral oil. The DFT method (B3LYP) at the 6-31G* basis set is used as before. The definition of the *IP* in liquid phase can be defined as:

$$IP = E^{A^+} - E^A + V_0, \tag{2}$$

where V_0 is the energy of a quasi-free electron in a condensed state. The values of V_0 are generally much less than the *IP* of the molecules (<0.1 eV). Hence, the V_0 is neglected in this work [30].

The electronic properties of molecules will be changed under the external field, resulting in changes of the ionization characteristic [31,32]. The electric field intensity of streamer initiation is about 10^8 V/m [33–35]. However, the electric field intensity of streamer fast propagation is at least 10^9 V/m [19]. The external field directions include X, X-, Y, Y-, Z and Z-. Electronic properties of a triolein molecule under an external electric field were investigated in our previous study and the direction of the electric field is along the Z axis [31].

Koopman's theorem points out that there is a negative correlation between the IP and the energy of the highest occupied molecular orbital (HOMO). The increase of HOMO energy means that the electrons easily escape from the molecular orbital and the insulation capacity declines. The HOMO energies of those molecules under an external field are calculated. Under an external electric field, the Hamiltonian of the molecule system is the sum of molecule Hamiltonian and interaction Hamiltonian, described as [36]:

$$H = H_0 + H_{int} \tag{3}$$

The H_{int} is the interaction Hamiltonian between the electric field and molecule, which can be described as:

$$H_{int} = -\boldsymbol{\mu} \cdot \boldsymbol{F} \tag{4}$$

where $\boldsymbol{\mu}$ is dipole moment and $\boldsymbol{F}$ is radiation field. External electric fields are applied along the X, X-, Y, Y-, Z and Z-axis. The intensities are 0.0005 a.u. and 0.005 a.u. 1 a.u. of electric field is equal to 5.14×10^{11} V/m. The intensities are namely 2.57×10^8 V/m and 2.57×10^9 V/m.

3. Results and Discussion

3.1. Wavenumber of Functional Group

The infrared spectrum reflects the molecular structure. However, the molecular structure determines the physicochemical and biological properties of the molecule. Wavenumbers of functional groups of the molecules are calculated. A scale factor of 0.96 for B3LYP/6-31G* is used to address the fundamental error [37]. The results are listed in Table 2 together with the experimental data published previously [38–40]. The spectral peak of a wavenumber calculated for 1-methylnaphthalene molecule is 3072.9 cm^{-1} from C–H stretching on benzene ring. The spectral peak of a wavenumber calculated for eicosane molecule is 3072.9 cm^{-1} from C–H stretching. For a triolein molecule, the wavenumbers of C=O stretching are 1775.7 cm^{-1}, 1778.2 cm^{-1} and 1784.9 cm^{-1} from the three C=O double bonds; the wavenumbers of C=C stretching are 1673.4 cm^{-1}, 1673.5 cm^{-1} and 1673.6 cm^{-1} from the three C=C double bonds. For a tristearin molecule, the wavenumbers of C=O stretching are 1775.6 cm^{-1}, 1778.2 cm^{-1} and 1784.8 cm^{-1} from the three C=O double bonds. The experimental values of wavenumbers are also listed in this table, which show consistencies with theoretical results.

Table 2. Wavenumber of functional groups.

Typical Molecule	Vibration Mode	Wavenumber (cm^{-1})	
		Calculated Value	Experimental Value
1-Methylnaphthalene	C–H stretching	3072.9	3077 [1]
Eicosane	C–H stretching	2954.9	2934 [1]
Triolein	C=O stretching	1775.7, 1778.2, 1784.9	1746 [2], 1745 [3]
	C=C stretching	1673.4, 1673.5, 1673.6	1653 [2], 1654 [3]
Tristearin	C=O stretching	1775.6, 1778.2, 1784.8	

[1,2,3] Experimental values are taken from [38–40], respectively.

3.2. Ionization Potentials and HOMO

The IPs of the typical molecules are calculated by the methods mentioned in Section 2 and the results are listed in Table 3. As shown in the table, the liquid-phase calculations are smaller than the gas-phase values. This is ascribed to the fact that the energy of a neutral molecule E^A is almost unchanged and the energy of a positive ion E^{A+} is decreased in PCM. The IP of a triolein molecule in liquid phase is 6.65 eV, which is comparable to that of a 1-methylnaphthalene molecule. The IP of a tristearin molecule in liquid phase is 7.73 eV, which is near to that of eicosane molecule. The calculated value of 1-methylnaphthalene is close to the experimental value, which assures the reliability of our method.

Table 3. Ionization potentials of typical molecules.

Typical Molecule	IP in Gas Phase (eV)	IP in Liquid Phase (eV)
1-Methylnaphthalene	7.51	6.25/6.20 [1]
Eicosane	8.78	7.92
Triolein	7.24	6.65
Tristearin	8.21	7.73

[1] The experimental value in Ref [41].

Electrons move on the HOMO of a molecule and may escape from the orbital by way of the energy obtained from collision, illumination or otherwise [33,42]. Collisions occur between neutral molecules and free electrons accelerated in the high electric field. A neutral molecule is ionized during the collision. A neutral molecule can also be ionized under illumination, namely photo-ionization. Table 4 shows the HOMO energies of typical molecules. The HOMO energies of typical molecules in the gas phase are -5.67 eV, -7.61 eV, -6.28 eV and -7.32 eV, respectively. The liquid-phase results are almost equal to the gas-phase ones. As discussed in Section 2, in PCM, the surrounding liquid of the molecule is considered as a continuum dielectric with a dielectric constant ε. The PCM does not much affect the electron structure of a neutral molecule.

Table 4. Highest occupied molecular orbital (HOMO) energies of typical molecules.

Typical Molecule	HOMO in Gas Phase (eV)	HOMO in Liquid Phase (eV)
1-Methylnaphthalene	-5.67	-5.71
Eicosane	-7.61	-7.60
Triolein	-6.28	-6.33
Tristearin	-7.32	-7.46

The isosurface of the HOMO of a 1-methylnaphthalene molecule is shown in Figure 2. The HOMO compositions are analyzed by the Mulliken method using the Multiwfn program [43]. C10, C11, C5 and C2 make the primary contribution to the HOMO with the percentage of 18.33%, 17.23%, 15.12% and 14.99%, respectively. Furthermore, the contribution of each C atom consists of two P-shells (10.87% and 7.41%, take C10 for example). It can be speculated that the electrons of these atoms may escape from the HOMO of 1-methylnaphthalene. The contributions of all the H atoms are all less than 1.5%. The H atoms are least likely to provide electronic in the discharge process.

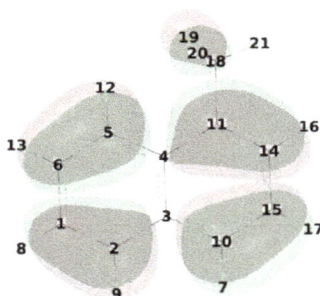

Figure 2. Isosurface of HOMO of 1-methylnaphthalene molecule.

The isosurface of the HOMO of an eicosane molecule is shown in Figure 3 and only H's labels are excluded in order to be concise. HOMO compositions are distributed across all the C atoms. In the order of the most contribution, it goes: C29 with 8.15%, C32 with 8.15%, C26 with 7.88%, C35 with 7.87%, C23 with 7.35%, C38 with 7.36%, C20 with 6.61%, C41 with 6.61%, C17 with 5.71%, C44 with 5.71%, C14 with 4.70%, C47 with 4.70%, C11 with 3.63%, C50 with 3.64%, C8 with 2.70%, C53 with 2.70%, C5 with 1.73%, C56 with 1.73%, C1 with 1.05% and C57 with 1.05%, respectively. From the results listed above, the C atoms in the molecule center are more likely to lose the electrons than the C atoms in two-ends.

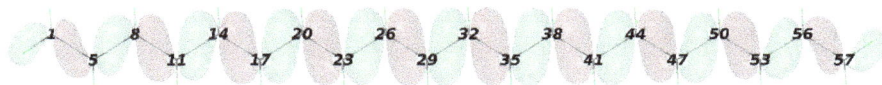

Figure 3. Isosurface of HOMO of eicosane molecule.

Figure 4 shows the isosurface of the HOMO of a triolein molecule. HOMO compositions are contributed by the C atoms in the *cis* C=C functional group at the position Sn-2. C100 and C102 make the primary contribution to the HOMO with the percentage of 39.67% and 39.66%, respectively. The contributions from the two C atoms consist of two P-shells (about 22.46% and 16.92% for C100; 22.45% and 16.92% for C102). The *cis* C=C functional group makes the triolein molecule very active. The molecule is easily ionized under a high electrical field. The IP of a triolein molecule is comparable to that of a 1-methylnaphthalene molecule.

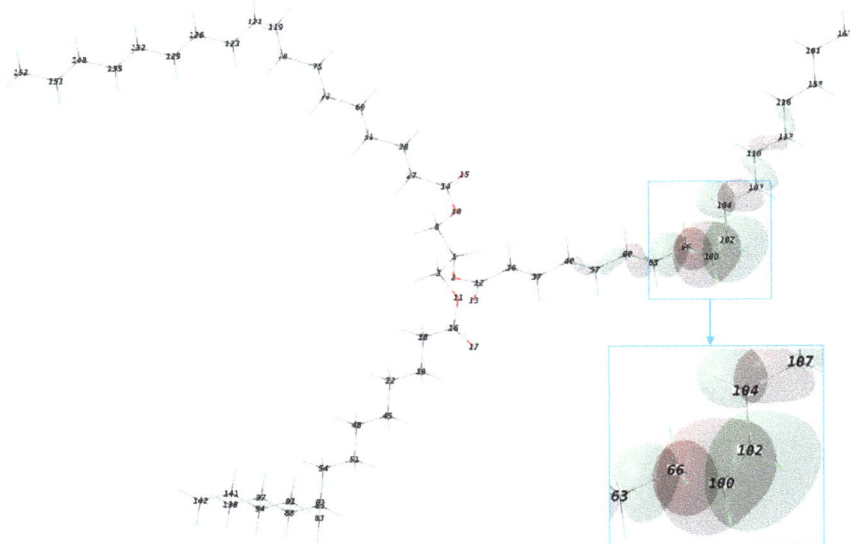

Figure 4. Isosurface of HOMO of triolein molecule.

The isosurface of the HOMO of a tristearin molecule is shown in Figure 5. O17 and the neighboring atom C18 make the primary contribution with a percentage of 68.42% and 10.62%, respectively. The contribution of O17 consists of two P-shells (about 43.60% and 24.83%). Compared with the *cis* C=C, the C=O has improved chemical stability. The IP of a tristearin molecule is comparable to that of an eicosane molecule.

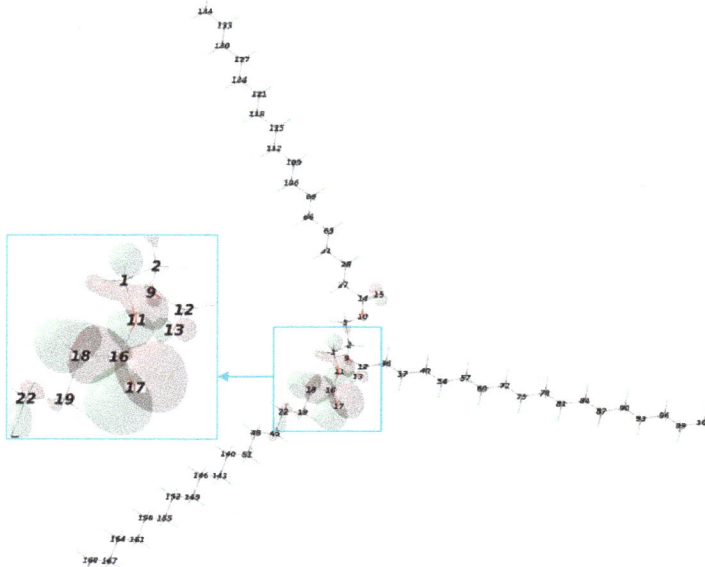

Figure 5. Isosurface of HOMO of tristearin molecule.

3.3. HOMO Variation under External Electric Field

The electronic structure of a molecule is not invariable under an external electric field as discussed in Section 2. The HOMO energy variations of the four typical molecules with applied fields are shown in Figure 6. The electric field intensities are 2.57×10^8 V/m and 2.57×10^9 V/m. The external field directions include X, X-, Y, Y-, Z and Z- of the Cartesian axis as shown in Figure 1. The HOMO energies of the four molecules have slight variations in all directions when the electric field intensity is 2.57×10^8 V/m, as shown in Figure 6. When the electric field intensity is 2.57×10^9 V/m, the HOMO energies change obviously. It is noticed that the HOMO energies of the two triglyceride molecules, including the triolein molecule and the tristearin molecule, increase sharply in all directions. There is a negative correlation between the IP and the energy of the HOMO. That means that the insulation characteristics of triolein and tristearin are more likely to be degraded under an external electric field than those of 1-methylnaphthalene and eicosane.

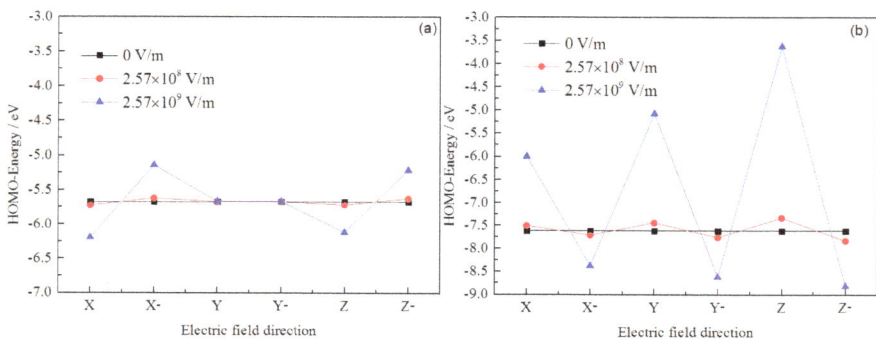

Figure 6. *Cont.*

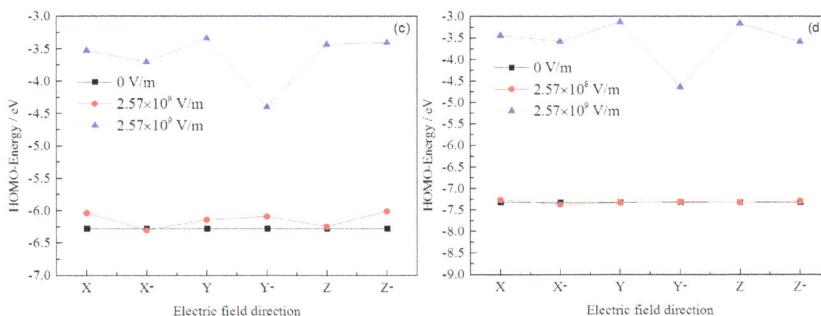

Figure 6. HOMO energy variations with applied fields: (**a**) 1-Methylnaphthalene; (**b**) eicosane; (**c**) triolein; and (**d**) tristearin.

Performance degradation of insulating material is affected by many factors, such as impurity, moisture or ageing [44,45]. Under the external electric fields, the variations of HOMO energy can be ascribed to the variations of composition of the HOMO. Figure 7 shows the isosurface of HOMO of molecule under the external electric field along X-axis with the intensity of 2.57×10^9 V/m. For 1-Methylnaphthalene as shown in Figure 7a, C10, C11, C5 and C2 also make the primary contribution to the HOMO, which are similar with the molecule under no external electric field as shown in Figure 2. However, the contribution rates of C5 and C2 increased to 16.81% and 16.77%, respectively. For triolein, as shown in Figure 7c, C81 and C83 make the primary contribution to the HOMO with the percentage of 40.12% and 34.04%, respectively. Comparison of Figures 4 and 7c, it is noticed that the total contribution rate of atoms of *cis* C=C decreases and the contribution rate of the atoms in the end of the carbon chain increases under the external electric field.

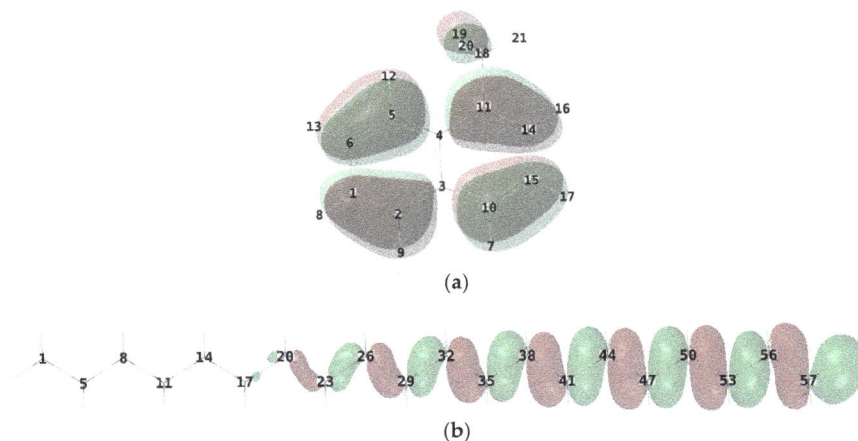

(**a**)

(**b**)

Figure 7. *Cont.*

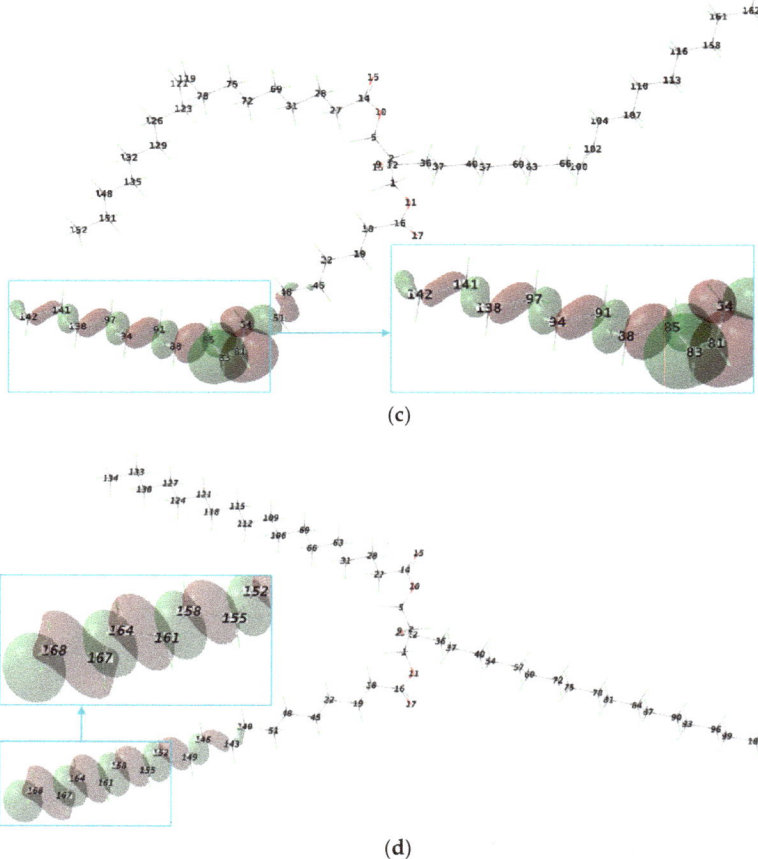

Figure 7. Isosurface of HOMO of molecule under the external electric field along X-axis with the intensity of 2.57×10^9 V/m: (**a**) 1-methylnaphthalene; (**b**) eicosane; (**c**) triolein; and (**d**) tristearin.

4. Discharge Mechanisms

4.1. Streamer Characteristic Effected by the Distribution of Ionization Potential

Each oil is considered in terms of a low IP molecules group and a high IP molecules group. The corresponding number densities within each group are N_L and N_H, respectively. For vegetable oil, the low IP molecules group is formed by triolein molecules and the high IP molecules group is composed of tristearin molecules. It is found that $N_L \gg N_H$ in vegetable oil. For mineral oil, the low IP molecules group is represented by 1-methylnaphthalene molecules and the high IP molecules group contains eicosane molecules. It is recognized that $N_L \ll N_H$ in mineral oil. Additionally, the IP of the low IP molecules group in vegetable oil is comparable to the IP of the low IP molecules group in mineral oil.

Consider the positive streamer in oils. When enough voltage is applied to the needle electrode, remarkable ionization of molecules happens. The generated electrons rush to the positive electrode quickly due to the high electron mobility and the positive ions are left to form the tip of the streamer. The generated positive ions move to the negative electrode slowly due to the low ion mobility.

The separation of positive ion and electron will form the induced electric field E, which can be written as:

$$\nabla \cdot E = \sum_i e n_i / \varepsilon, \tag{5}$$

where n_i is the number density of the molecule i which has been ionized and ε is the dielectric constant. Assume that all the low IP molecules are ionized when breakdown happens. Equation (5) can be revised as:

$$\nabla \cdot E = e N_L / \varepsilon. \tag{6}$$

Due to $N_L >> N_H$ in vegetable oil, there will be sufficient low IP molecules being ionized that form the powerful induced electric field E, which facilitates the further ionization of molecules. The streamer velocities of vegetable oils can reach or exceed 10 km/s. Due to $N_L << N_H$ in mineral oil, the limited low IP molecules can be ionized to form the weak induced electric field E. The streamer velocities of mineral oil remain quite slow as 1–2 km/s. In order to observe the fast streamer in mineral oil, a higher voltage is needed after breakdown happens. Assume that all the high IP molecules are ionized when a fast streamer appears in mineral oil. Sufficient high IP molecules can be ionized to form a powerful induced electric field E. Equation (5) can be revised as:

$$\nabla \cdot E = e(N_L + N_H) / \varepsilon. \tag{7}$$

4.2. Polarity Effect and Space Charge

In Reference [15], the breakdown voltage of vegetable oil is lower than that of mineral oil both under positive polarity and under negative polarity. However, it is worth pointing out that the streamer velocity of vegetable oil can reach 10~30 km/s (fast streamer) under positive polarity but be severely suppressed under negative polarity. The streamer velocity of mineral oil is even higher than that of vegetable oil under negative polarity. It is almost impossible that the streamer velocity of mineral oil is higher than that of vegetable oil under positive polarity. Probably because the number density of low IP molecules group in vegetable oil is much greater than that in mineral oil, that is $N_{LV} >> N_{LM}$, the breakdown voltage of vegetable oil is always lower than that of mineral oil.

Also, due to $N_{LV} >> N_{LM}$, the polarity effect in vegetable oil is more obvious than that in mineral oil. This eventually leads to the different streamer characteristics in vegetable oil and mineral oil under positive polarity and negative polarity. Figure 8 shows the polarity effect in vegetable oil and mineral oil.

Under positive polarity in Figure 8a, the electrons enter the positive needle electrode after the neutral molecules are ionized. The positive ions are left near the needle tip. The space charges weaken the field near the tip and strengthen the field forward. The effects of space charges on the electric field in vegetable oil are more significant than that in mineral oil. The strengthened field facilitates the further ionization process. The induced electric field will dominate the streamer propagation in the long oil gaps. The insulation degradation of molecules in such high fields will further facilitate the streamer propagation. Hence, the fast streamer appears easily in vegetable oils.

Under negative polarity in Figure 8b, the electrons rush to the positive plane electrode after the neutral molecules are ionized. The positive ions are left behind. The space charges strengthen the field near the tip and weaken the field forward. The effects of space charges on field in vegetable oil are also more significant than that in mineral oil. The weakened field represses the further ionization process. Hence, the streamer velocities of mineral oil even surpass those of vegetable oil in long oil gaps.

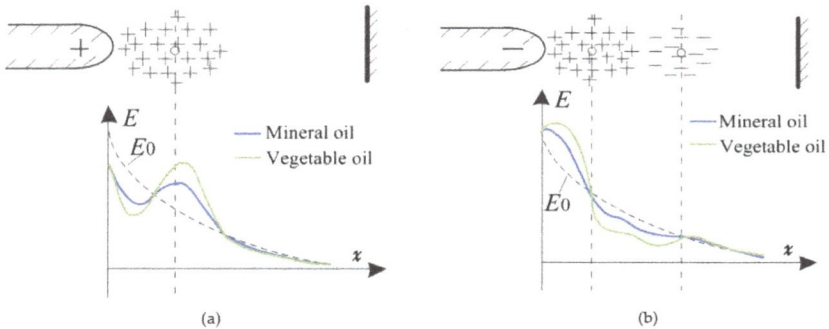

Figure 8. Polarity effect in vegetable oil and mineral oil: (**a**) positive polarity; and (**b**) negative polarity.

5. Conclusions

In this work, 1-methylnaphthalene and eicosane are selected as the typical molecules in mineral oil. Triolein and tristearin are selected as the typical molecules in vegetable oil. The electronic properties of the typical molecules are calculated using the quantum chemistry method (B3LYP/6-31G*) based on density functional theory. The mechanisms of fast and slow streamers in mineral oil and vegetable oil are discussed based on the calculations. Several conclusions are summarized as follows:

(1).   The IP of a triolein molecule is comparable to that of a 1-methylnaphthalene molecule. The IP of a tristearin molecule is comparable to that of an eicosane molecule, especially in the liquid phase. The mechanisms of losing electrons are discussed based on the analysis of HOMO composition.

(2).   The HOMO energy variations with applied fields are calculated. The insulation characteristics of triolein and tristearin are more likely to be degraded under an external electric field than those of the 1-methylnaphthalene and eicosane.

(3).   Due to the fact that the number density of the low IP molecules group in vegetable oil is much greater than that in mineral oil, the polarity effect in vegetable oil is more obvious than that in mineral oil. This eventually leads to different streamer characteristics in vegetable oil and mineral oil under positive polarity and negative polarity.

Acknowledgments: The work is supported by the National Natural Science Foundation of China (No. 51425702). The authors appreciate the National "111" Project of the Ministry of Education of China (No. B08036) and the State Key Program of National Natural Science Foundation of China (No. U1537211). The computational resources and Gaussian program are provided by National Supercomputing Center in Shenzhen (Gaussian 09 D01: TCP-Linda). We also thank to Special Program for Applied Research on Super Computation of the NSFC-Guangdong Joint Fund (the second phase).

Author Contributions: Feipeng Wang and Jian Li structured this work. Yachao Wang and Jinghan Zhou conducted the molecule simulation. Yachao Wang and Suning Liang analyzed the data and wrote the paper, Feipeng Wang and Jian Li revised the manuscript.

Conflicts of Interest: The authors declare no conflict of interest.

References

1.   Jing, Y.; Timoshkin, V.; Given, M.J.; Macgregor, S.G.; Wilson, M.P.; Wang, T. Dielectric properties of natural ester, synthetic ester Midel 7131 and mineral oil Diala D. In *Proceedings of the Power Modulator and High Voltage Conference*, San Diego, CA, USA, 3–7 June 2012; pp. 63–66.

2.   Fernández, I.; Delgado, F.; Ortiz, F.; Ortiz, A.; Fernández, C.; Renedo, C.J.; Santisteban, A. Thermal degradation assessment of Kraft paper in power transformers insulated with natural esters. *Appl. Therm. Eng.* **2016**, *104*, 129–138. [CrossRef]

3.   Xiang, C.; Zhou, Q.; Li, J.; Huang, Q.; Song, H.; Zhang, Z. Comparison of dissolved gases in mineral and vegetable insulating oils under typical electrical and thermal faults. *Energies* **2016**, *9*, 312. [CrossRef]

4. Lashbrook, M.; Gyore, A.; Martin, R. A review of the fundamental dielectric characteristics of ester-based dielectric liquids. *Proc. Eng.* **2017**, *202*, 121–129. [CrossRef]

5. Oommen, T.V.; Claiborne, C.C.; Mullen, J.T. Biodegradable electrical insulation fluids. In Proceedings of the Electrical Insulation Conference and Electrical Manufacturing and Coil Winding Conference, Rosemont, IL, USA, 25–25 September 1997; pp. 465–468.

6. Zou, P.; Li, J.; Sun, C.; Liao, R.; Zhang, Z. Influences of moisture content on insulation properties of vegetable insulating oil. *High Vol. Eng.* **2017**, *37*, 1627–1633.

7. Liao, R.; Hao, J.; Chen, G.; Ma, Z.; Yang, L. A comparative study of physicochemical, dielectric and thermal properties of pressboard insulation impregnated with natural ester and mineral oil. *IEEE Trans. Dielectr. Electr. Insul.* **2011**, *18*, 1626–1637. [CrossRef]

8. Cargill. Envirotemp FR3 Fluid. Available online: https://www.cargill.com/bioindustrial/envirotemp/fr3 (accessed on 1 November 2017).

9. Kolcunová, I.; Kurimský, J.; Cimbala, R.; Petráš, J.; Dolník, B.; Džmura, J.; Balogh, J. Contribution to static electrification of mineral oils and natural esters. *J. Electrost.* **2017**, *88*, 60–64. [CrossRef]

10. Pourrahimi, A.M.; Olsson, R.T.; Hedenqvist, M.S. The role of interfaces in polyethylene/metal-oxide nanocomposites for ultrahigh-voltage insulating materials. *Adv. Mater.* **2018**, *30*, 1703624. [CrossRef] [PubMed]

11. Aljurea, M.; Becerraa, M.; Pallon, L.K.H. Electrical conduction currents of a mineral oil-based nanofluid in needle-plane configuration. In Proceedings of the IEEE Conference on Electrical Insulation and Dielectric Phenomena, Toronto, ON, Canada, 16–19 October 2016; pp. 687–690.

12. Nilsson, F.; Karlsson, M.; Pallon, L.; Giacinti, M.; Olsson, R.T.; Venturi, D.; Gedde, U.W.; Hedenqvist, M.S. Influence of water uptake on the electrical DC-conductivity of insulating LDPE/MgO nanocomposites. *Compos. Sci. Technol.* **2017**, *152*, 11–19. [CrossRef]

13. Pourrahimi, A.M.; Pallon, L.K.H.; Liu, D.; Hoang, T.A.; Gubanski, S.; Hedenqvist, M.S.; Olsson, R.T.; Gedde, U.W. Polyethylene nanocomposites for the next generation of ultralow transmission-loss HVDC cables: Insulation containing moistureresistant MgO nanoparticles. *ACS Appl. Mater. Interfaces* **2016**, *8*, 14824–14835. [CrossRef] [PubMed]

14. Fofana, I.; Borsi, H.; Gockenbach, E. Fundamental investigations on some transformer liquids under various outdoor conditions. *IEEE Electr. Insul. Mag.* **2001**, *8*, 1040–1047. [CrossRef]

15. Liu, Q.; Wang, Z.D. Streamer characteristic and breakdown in synthetic and natural ester transformer liquids under standard lightning impulse voltage. *IEEE Trans. Dielectr. Electr. Insul.* **2011**, *18*, 285–294. [CrossRef]

16. Kohn, W. Nobel Lecture: Electronic structure of matter—Wave functions and density functionals. *Rev. Mod. Phys.* **1999**, *71*, 1253–1266. [CrossRef]

17. Kumar, J.; Nemade, H.B. Adsorption of small molecules on niobiumdoped graphene: A study based on Density Functional Theory. *IEEE Electron Device Lett.* **2018**, *39*, 296–299. [CrossRef]

18. Wedin, P. Electrical breakdown in dielectric liquids-a short overview. *IEEE Electr. Insul. Mag.* **2014**, *30*, 20–25. [CrossRef]

19. Rozga, P. Streamer propagation in small gaps of synthetic ester and mineral oil under lightning impulse. *IEEE Trans. Dielectr. Electr. Insul.* **2015**, *22*, 2754–2762. [CrossRef]

20. Claiborne, C.C.; Pearce, H.A. Transformer fluids. *IEEE Electr. Insul. Mag.* **1989**, *5*, 16–19. [CrossRef]

21. Wang, X.; Zeng, Q.; Verardo, V.; Contreras, M.d.M. Fatty acid and sterol composition of tea seed oils: Their comparison by the "FancyTiles" approach. *Food Chem.* **2017**, *233*, 302–310. [CrossRef] [PubMed]

22. Fatemi, S.H.; Hammond, E.G. Glyceride structure variation in soybean varieties. I. Stereospecific analysis. *Lipids* **1977**, *12*, 1032–1036. [CrossRef]

23. Lísa, M.; Holčapek, M. Triacylglycerols profiling in plant oils important in food industry, dietetics and cosmetics using high-performance liquid chromatography–atmospheric pressure chemical ionization mass spectrometry. *J. Chromatogr. A* **2008**, *1198–1199*, 115–130. [CrossRef] [PubMed]

24. Takagi, T. Stereospecific analysis of triacyl-sn-glycerols by chiral high-performance liquid chromatography. *Lipids* **1991**, *26*, 542–547. [CrossRef]

25. Zeb, A. Triacylglycerols composition, oxidation and oxidation compounds in camellia oil using liquid chromatography–mass spectrometry. *Chem. Phys. Lipids* **2012**, *165*, 608–614. [CrossRef] [PubMed]

26. Frisch, M.J.; Trucks, G.W.; Schlegel, H.B.; Scuseria, G.E.; Robb, M.A.; Cheeseman, J.R.; Scalmani, G.; Barone, V.; Mennucci, B.; Petersson, G.A.; et al. *Gaussian 09*; Gaussian Inc.: Wallingford, CT, USA, 2013.

27. Becke, A.D. Density—Functional thermochemistry. III. The role of exact exchange. *J. Chem. Phys.* **1993**, *98*, 5648–5652. [CrossRef]

28. Lee, C.; Yang, W.; Parr, R.G. Development of the Colle-Salvetti correlation-energy formula into a functional of the electron density. *Phys. Rev. B* **1988**, *37*, 785–789. [CrossRef]

29. Cossi, M.; Scalmani, G.; Rega, N.; Barone, V. New developments in the polarizable continuum model for quantum mechanical and classical calculations on molecules in solution. *J. Chem. Phys.* **2002**, *117*, 43–54. [CrossRef]

30. Ingebrigtsen, S.; Smalo, H.S.; Astrand, P.O.; Lundgaard, L.E. Effects of electron-attaching and electron-releasing additives on streamers in liquid cyclohexane. *IEEE Trans. Dielectr. Electr. Insul.* **2009**, *16*, 1524–1535. [CrossRef]

31. Wang, Y.; Wang, F.; Li, J.; Huang, Z.; Liang, S.; Zhou, J. Molecular structure and electronic properties of triolein molecule under an external electric field related to streamer initiation and propagation. *Energies* **2017**, *10*, 510. [CrossRef]

32. Xu, G.L.; Xie, H.X.; Yuan, W.; Zhang, X.Z.; Liu, Y.F. Properties of a Si_2N molecule under an external electric field. *Chin. Phys. B* **2012**, *21*, 053101. [CrossRef]

33. Tobazcon, R. Prebreakdown phenomena in dielectric liquids. *IEEE Trans. Dielectr. Electr. Insul.* **1994**, *1*, 1132–1147. [CrossRef]

34. Beroual, A.; Zahn, M.; Badent, A.; Kist, K.; Schwabe, A.J.; Yamashita, H.; Yamazawa, K.; Danikas, M.; Chadband, W.D.; Torshin, Y. Propagation and structure of streamers in liquid dielectrics. *IEEE Electr. Insul. Mag.* **1998**, *14*, 6–17. [CrossRef]

35. Badent, R.; Hemmer, M.; Konekamp, U.; Julliard, Y.; Schwab, A.J. Streamer inception field strengths in rape-seed oils. In Proceedings of the Annual Report Conference on Electrical Insulation and Dielectric Phenomena, Victoria, BC, Canada, 15–18 October 2000; pp. 272–275.

36. Buckingham, A.D. Direct method of measuring molecular quadrupole moments. *J. Chem. Phys.* **1959**, *30*, 1580–1585. [CrossRef]

37. The National Institute of Standards and Technology (NIST). Computational Chemistry Comparison and Benchmark Data Base. Available online: http://cccbdb.nist.gov/vibscalejust.asp (accessed on 31 October 2016).

38. The National Institute of Standards and Technology (NIST). NIST Chemistry Webbook. Available online: http://webbook.nist.gov/cgi/cbook.cgi?ID=C90120&Units=SI&Type=IR-SPEC&Index=2#IR-SPEC (accessed on 1 December 2017).

39. Christy, A.A.; Xu, Z.F.; Harrington, P.D.B. Thermal degradation and isomerisation kinetics of triolein studied by infrared spectrometry and GC–MS combined with chemometrics. Chem. Phys. *Lipids* **2009**, *158*, 22–31. [CrossRef] [PubMed]

40. Kos, A.; Tefelski, D.B.; Kosciesza, R.; Rościcki, A.J.; Roszkiewicz, A.; Ejchart, W.; Jastrzebski, C.; Siegoczyńsk, R.M. Certain physico-chemical properties of triolein and methyl alcohol–triolein mixture under pressure. *High Press. Res.* **2007**, *27*, 39–42. [CrossRef]

41. Holroyd, R.A.; Preses, J.M.; Boettcher, E.H.; Schmidt, W.F. Photoconductivity induced by single-photon excitation of aromatic molecules in liquid hydrocarbons. *Chem. Inf.* **1984**, *15*, 744–749. [CrossRef]

42. Sparks, M.; Mills, D.L.; Warren, R.; Holstein, T.; Maradudin, A.A.; Sham, L.J.; Loh, J.E.; King, D.F. Theory of electron-avalanche in solids. *Phys. Rev. B* **1981**, *24*, 3519–3536. [CrossRef]

43. Lu, T.; Chen, F.W. *Multiwfn 3.3.8*; Kein Research Center for Natural Sciences: Beijing, China, 2016.

44. Akhlaghi, S.; Pourrahimi, A.M.; Hedenqvist, M.S.; Sjöstedt, C.; Bellander, M.; Gedde, U.W. Degradation of carbon-black-filled acrylonitrile butadiene rubber in alternative fuels: Transesterified and hydrotreated vegetable oils. *Polym. Degrad. Stab.* **2016**, *123*, 69–79. [CrossRef]

45. Akhlaghi, S.; Pourrahimi, A.M.; Sjöstedt, C.; Bellander, M.; Hedenqvist, M.S.; Gedde, U.W. Degradation of fluoroelastomers in rapeseed biodiesel at different oxygen concentrations. *Polym. Degrad. Stab.* **2017**, *136*, 10–19. [CrossRef]

MDPI

St. Alban-Anlage 66

4052 Basel

Switzerland

Tel. +41 61 683 77 34

Fax +41 61 302 89 18

www.mdpi.com

Energies Editorial Office

E-mail: energies@mdpi.com

www.mdpi.com/journal/energies

www.ingramcontent.com/pod-product-compliance
Lightning Source LLC
Chambersburg PA
CBHW041216220326
41597CB00033BA/5985